IMAGES
of America

FORT CARSON

map of fort carson

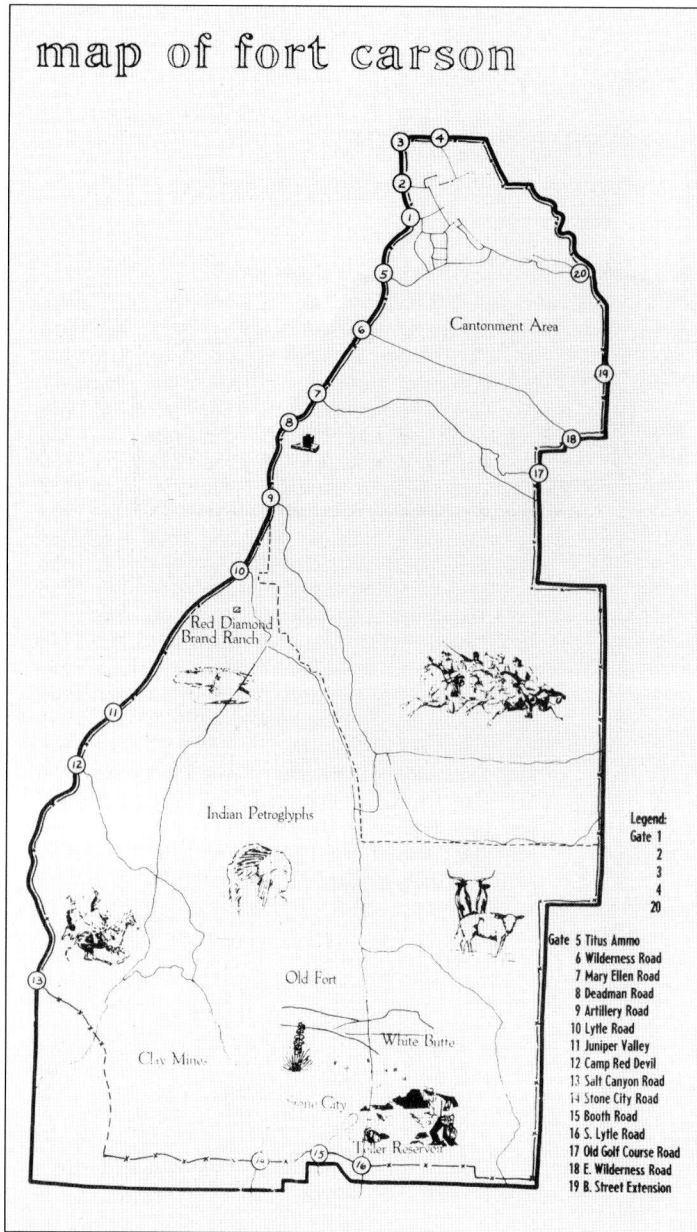

Cantonment Area

Red Diamond
Brand Ranch

Indian Petroglyphs

Old Fort

Clay Mines

White Butte

Stone City

Teller Reservoir

Legend:
Gate 1
    2
    3
    4
    20

Gate 5 Titus Ammo
   6 Wilderness Road
   7 Mary Ellen Road
   8 Deadman Road
   9 Artillery Road
  10 Lytle Road
  11 Juniper Valley
  12 Camp Red Devil
  13 Salt Canyon Road
  14 Stone City Road
  15 Booth Road
  16 S. Lytle Road
  17 Old Golf Course Road
  18 E. Wilderness Road
  19 B. Street Extension

All 20 installation gates are marked on this rough outline of the Fort Carson military installation boundaries in 1970. Other landmarks mentioned in this book are identified as well: Dead Man's Canyon, the Indian rock art area, the clay mines, Stone City, and Teller Reservoir (Courtesy of the Department of Defense.)

ON THE COVER: White military buildings sit at the foot of Cheyenne Mountain, the ever-present backdrop to Fort Carson activities and the source of the facility's nickname—the Mountain Post. This image represents Carson's transition from camp to fort, and from pack to mechanized, when bubble helicopters replaced the beloved pack mules. Such is the nature of an Army installation that constantly modernizes with changes in warfare. (Courtesy of the 4th Infantry Division Museum.)

IMAGES
of America

FORT CARSON

Angela Thaden Hahn and Joseph E. Berg

ARCADIA
PUBLISHING

Published by Arcadia Publishing
Charleston, South Carolina

Printed in the United States of America

Library of Congress Control Number: 2019954259

For all general information, please contact Arcadia Publishing:
Telephone 843-853-2070
Fax 843-853-0044
E-mail sales@arcadiapublishing.com
For customer service and orders:
Toll-Free 1-888-313-2665

Visit us on the Internet at www.arcadiapublishing.com

*To the thousands of American soldiers and families
who have called Fort Carson home.*

CONTENTS

ACKNOWLEDGMENTS

Many thanks to Mike Kline of the 4th Infantry Division Museum; Jennifer Kolise and James Kulbeth of the Cultural Resources Program; Jim Drye of the 4th Squadron, 12th Cavalry Horse Platoon; Mike Hahn for technical military input and steady support; Nancy Thaden and Tony Delphia for edits; and the many Department of Defense photographers from the 1940s to the present.

Unless otherwise noted, all images appear courtesy of the 4th Infantry Division Museum.

INTRODUCTION

Most US Army posts are named after generals made famous by acts of heroism or noteworthy service. Likewise, Camp Carson was named for soldier Christopher "Kit" Houston Carson (1809–1868). While hunting and trapping all over the western United States, he spent some time in Pueblo and at the trading post known as Bent's Fort on the Arkansas River. After meeting explorer John C. Fremont, Carson turned to guiding and scouting across Colorado's Rocky Mountains. He served as commanding general of Fort Garland near Alamosa and retired near Las Animas, downriver from the Piñon Canyon Maneuver Site frequented by thousands of Fort Carson troops.

Long before Kit Carson, or any other human came to the area, the land lay underwater, as is evidenced by pieces of petrified wood and abundant fossilized sea life scattered over the reservation's open spaces. Once the seas subsided, dinosaurs passed through, leaving footprints in the mud. Later, Indigenous peoples wandered into the ancient region watered by a stream now known as Turkey Creek. They hunted turkey and deer and camped under rock shelters in rooms hollowed out by erosion of the Dakota sandstone. They left traces of their life in the form of rock art and groups of vertical grooves in the rock walls where they sharpened their bladed tools. Bedrock metates reveal where they ground dried corn and berries with a stone into something edible.

In the more recent history of the region, an early visitor came in 1806 when Lt. Zebulon Pike set out under the direction of Pres. Thomas Jefferson to explore the southern reaches of the newly acquired Louisiana Purchase. Having arrived at Pueblo in early November, Pike and his men decided to ascend the tallest peak to their northwest. They headed in a northerly direction up the Fontaine que Bouille, now known as Fountain Creek. Some researchers say the men parted from the creek at the Little Fontaine que Bouille and headed northwest toward the mountains. This would have taken them right across the present-day Fort Carson range. They made it to the summit of Mount Rosa but found their destination was not as close as it had appeared down below. Pike decided they were ill-prepared for the waist-deep snow they encountered. Equipped with little food, poor footwear, and lightweight clothing, the young lieutenant determined it was impossible for anyone to ever get to the top of the grand peak. Nevertheless, the mountain was later named for him.

Meanwhile, several Indian tribes continued to use the land for hunting and camping. Ute, Comanche, Kiowa, Cheyenne, Arapaho, Sioux, Jicarilla Apache, and Pawnee all passed through the region frequently. These groups left behind arrowheads, pottery fragments, and rock shards from making tools.

After an 1861 treaty, many of the Cheyenne and Arapaho left the area, which became part of the new Colorado Territory. White settlers slowly began to trickle into the creek drainages, but clashes between Indians and settlers occurred well into the 1860s. At the mention of an Indian scare, settlers hustled off to one of the few fortified homes located between the nearest towns of Colorado City to the north and Fountain to the east. By 1869, the area was patrolled by hundreds of cavalrymen, and most remaining Indians then left.

It was in 1863 when Henry Harkens was found dead at his cabin by his two partners. He had been living near the upper Little Fountain Creek in Sawmill Canyon. Judging by the hatchet wounds in Harkens's body, his friends quickly determined he had been attacked by Indians. However, the sheriff and a posse from Fountain found tracks in their investigation that led to the murderous gang known as the Espinosas. These outlaws had ridden the trail up from Canon City, now known as Highway 115, on a quest to kill as many gringos as they could find. They managed to slay 30 to 40 settlers before they were stopped. Harkens was laid to rest on top of a knoll overlooking the gulch that now goes by the name of Dead Man's Canyon.

A look at original plat surveys taken from 1863 to 1877 of the land that is now Fort Carson reveals some interesting insights. Cut through the piñon and juniper trees were several roads used for hauling wood, as well as numerous other roads that joined with the Canon City Road. Very few homesteads are marked on these surveys, suggesting the region was sparsely populated. The southeast portion of the present Fort Carson area is lined with dry arroyos and afforded precious little water to farmers and ranchers. The maps show that most settlers, including a sheep rancher, chose land on the west side along Rock Creek, Little Fountain Creek, Red Creek, and Turkey Creek.

Settlers continued to move into the area, leaving traces of their existence in some form or another. A well-known landmark among soldiers on maneuvers called Booth Mountain bears the name of an early rancher, ensuring he will never be forgotten. It appears as a high, grey outcropping of rounded boulders. Gold-mining tycoon Spencer Penrose left his mark along Turkey Creek when he built a charming weekend bungalow. Most area residents lived in crude log cabins. Indeed, early censuses show that enough families were living in the area to justify at least one schoolhouse and a post office in Lytle, a small community on Turkey Creek that served the local people until about 1920.

On the eastern side of the region, cattle roamed on the rugged, open range replete with dry sand and clay, both of which fly as fine dust in high winds. Characteristic to this region are rounded granite boulders of varying sizes, spear-like yucca, cholla cactus, and golden rabbitbrush.

The reservation boundaries eventually expanded to include the ghost town of Stone City to the south, which began as a mining town for the workers in nearby stone quarries and clay mines. About 100 residents lived in the company town and had access to a general store, post office, school, and a music teacher. The caves are now home to protected bats.

Efforts have been made by Fort Carson's Cultural Resources department to discover, collect, study, and preserve the rich history of the land now occupied by one of the most successful Army posts in the country.

Following Japan's attack on Pearl Harbor in 1941, many Army camps sprung up across the country for the purpose of training soldiers for the war. At the same time, hundreds of vacant homes spelled stagnant growth for Colorado Springs. Local businessmen and politicians convinced the War Department that the Pikes Peak region had miles of prairie suitable for large-scale training maneuvers and a favorable climate to permit training year-round. The varied terrain offered many opportunities for soldiers to learn and practice commando skills. Subsequently, the city bought a ranch south of town, and the government purchased many more. Floodlights were set up in January 1942, and construction hummed around the clock with a sense of urgency. Even severe winter weather and heavy spring rains did not deter the work.

The camp was designed to conform to the contour of the land to eliminate unnecessary grading, which created the unique banana shape so obvious on maps of the post. Slogans such as "Nail down the planks—here come the Yanks!" urged the workers to step up production, but it was a challenge to keep enough building material on hand. Lumber and pipe were in short supply due to many other military camps being constructed at the same time across the country. Interestingly, permanency was not a factor, as the structures were built with only a five-year life expectancy in mind. Most of the country's Army camps thrown up at this time had the uniform look of rectangular, two-story, wooden, plank-sided buildings, painted white, and lined up in neat rows. All construction at Camp Carson was completed by November 1942.

The new post became the training grounds for numerous Army divisions and various smaller units such as tank battalions, a Greek infantry battalion, and an Italian ordnance company. By late 1943, there were 43,000 military personnel stationed at Camp Carson to learn their military occupation specialty (MOS), including nurses, cooks, replacement troops, and mule packers.

It comes as a surprise to many that an internment camp was opened in January 1943 in the northwestern corner of the reservation between Highway 115 and the post rail yard. The compound was later designated a prisoner of war (POW) camp and eventually housed 3,000 German and Italian prisoners during World War II. An additional 5,000 prisoners were housed in overflow barracks east of Pershing Field. Besides working at local farms, prisoners were put to work in 17 other branch camps across the state. They worked in the post laundry and baking facilities and in their own canteen where the men purchased necessities. They kept busy in the prison woodshop and published their own newspaper. Carson's POW camp even had a cemetery for their countrymen who died while imprisoned. After the war, the bodies were shipped back to their homeland, and any remaining prisoners were returned to Europe in 1946.

When the war was over, activities at the camp were greatly reduced since the urgency for military training was over. The number of personnel at the camp dwindled to 600, and the installation was destined for closure. However, with the same tenacity that resulted in the initial creation of the post, the local populace persuaded the War Department in April 1946 to keep the camp open. More troops arrived with a regimental combat team and a pack battalion of famously stubborn Missouri-born mules.

In January 1950, tragedy struck Camp Carson when a smoldering fire at the Broadmoor golf course burst into flames in winds of 100 miles per hour. The fire spread rapidly southwest until it reached the northwest corner of the post and burned down the old POW camp and many of the depot warehouses. Nine soldiers and volunteer civilians died that day while fighting the fire, and about 30 were hospitalized. Survivors described how burning debris was pushed before the wind like fireballs and landed on rooftops, igniting more buildings. There were fires all over the post. The soldiers thought they would never get the fire under control. Nine streets on the post were renamed to honor those who died in the disaster.

During lighter times, actor Robert Mitchum came to Fort Carson in 1951 to star in the movie *One Minute to Zero*, about an Army officer in the early stages of the Korean War. The Mountain Post was chosen for filming because much of its terrain is like that found in Korea. Carson engineers built bridges, roads, and a 4,000-foot runway. Hundreds of Carson troops were used in the production. The set utilized the mock French village Beauclaire, which had been constructed years earlier for combat training. The site was transformed into straw, thatched huts, wet rice paddies, and muddy roads. After shooting the movie, the site became another realistic training setting for troops bound for Korea.

The temporary military camp withstood proposed closures and became an official fort in 1954. Although families started to arrive in the late 1940s, the first school did not open until September 1954 in an old World War II–era building, and the first permanent elementary school building was started in August 1956. Likewise, construction began in 1955 on quarters for 1,000 families.

Ten years later, when the Red Diamond Division occupied post headquarters, the reservation was enlarged to more than twice its size. As predicted, the post added to the city's economy by bringing in a population of military personnel and their families. Additionally, former Fort Carson troops often retired in the area. Several more times throughout the years, Carson became the discussion of possible closure. Just as housing was hard to find when the post first opened, houses became hard to sell in 1960, and promising bedroom communities like nearby Security saw an exodus of families as men were transferred out of Carson.

It was in June 1965 when the post came to the aid of the local community. Army helicopters evacuated thousands of civilians from a disastrous flood caused by days of heavy rain. Many victims were given medical care at the Army hospital and received food, blankets, and cots. Fort Carson property also suffered some damage when a 20-foot crest washed out the bridge on B Street just outside the gate. Additionally, two walls of water washed out 450 feet of track at the Kelker railroad siding. The flooding left the northern part of the fort covered with mud and silt.

Fort Carson became a busy wartime hub again in the mid-1960s with soldiers transferring back and forth from Viet Nam, but for the first time, community relations between the city and the post were low because of the growing nationwide anti-war movement. When the public was invited to participate in Armed Forces Day activities at Butts Field, anti-war protestors also showed up chanting, "Hell no, we won't go."

The Mountain Post was spared once again the threat of closure when the famed 4th Infantry Division relocated to Carson in 1970. Since then, the Mountain Post has been the home of the Ivy Division, with the exception of about 15 years. Author Angela Thaden Hahn moved to Fort Carson in 1970 with her father, who served in the Army as a supply sergeant. She walked to school across the street from her quarters and rode her bike all over the housing area to visit her friends. Her memories include freezing cold wind blowing hard enough to shatter the car windshield, knee-deep snow, fossils of seashells on the hill out back, horse-riding lessons at Turkey Creek Ranch, introduction to Indian culture at a powwow event on the post, and watching rappelling demonstrations at Cheyenne Canyon.

During the early 1970s, Carson was selected to test the concept of an all-volunteer Army. Many soldiers requested transfers to the Mountain Post for its favorable treatment of the men. Soldiers were regarded as individuals with ideas to share rather than as mere expendable tools of war. Many programs were implemented to make volunteering look attractive, thus maintaining a sizable force without conscription.

With advancements in warfare, an additional land deal of 235,300 acres made room for the Piñon Canyon Maneuver Site in September 1983. And although the Army has expressed interest in expanding the site with the purchase of even more ranches, the locals were not giving in so easily. The training site is located about 150 miles southeast of Fort Carson near the ranching community of Hohne. It is used for large force-on-force training and dismounted maneuvers. Rumbles from the big guns on Carson's gunnery range can be heard and felt for miles. These continual training exercises are necessary to hone the warfighting skills of Fort Carson soldiers.

For 75 years, the Mountain Post has remained a constant fixture in the Pikes Peak Region like the unchanging Cheyenne Mountain at its back door. The installation continues to be an ideal training ground for the most modern warfare and is described as the best hometown in the Army. Favorable relations between the Army and civilians remain strong as soldiers reach out in times of local disaster, and the citizenry responds with gratitude.

The city of Colorado Springs has long been appreciative of Fort Carson's presence. In 2000, the city presented the installation with a striking sculpture of the post's namesake, in recognition of the soldiers and families of Fort Carson who have given unselfish service to the nation, state, and community. Commanding Maj. Gen. Edward Soriano accepted the gift of the Kit Carson statue, which was placed at the front gate of the post to reflect the history of the region that is closely linked to the military. The general acknowledged the community's role in encouraging the preservation of the installation in the face of numerous budget cuts.

One

ANCIENT LIKE TURKEY CREEK

After the Civil War, soldier Christopher "Kit" Carson rose in rank to brigadier general and received command of Fort Garland, near Alamosa, in the San Luis Valley. He retired shortly thereafter and settled into a peaceful life of farming and ranching at Boggsville. His retirement home rested on the Purgatoire River, a few miles from where old Bent's Fort once flourished. Elbridge Ayer Burbank sketched this pencil drawing in 1863.

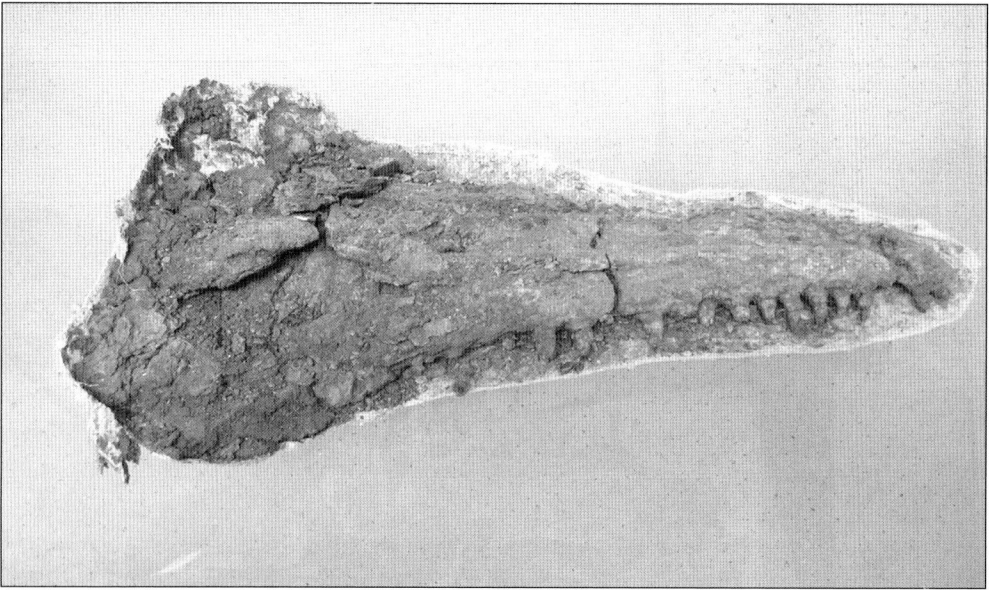

Discovered on the military reservation now known as Fort Carson, this eight-inch-long fossilized jawbone has become one of the thousands of artifacts in the large collection of Carson's Cultural Resources department. The many teeth along the lower edge suggest it belonged to a prehistoric eel. In the days before the Cultural Resources department came into existence, many interesting finds, such as this one, were collected without proper cataloging. (Angela Thaden Hahn.)

Cultural Resources keeps this illustration in its collection to show what a prehistoric eel may have looked like. A study of Colorado geology reveals that the high plains were covered multiple times by inland seas and lakes. Fossils of aquatic life found in layers of mudstone and limestone across vast areas, including the Fort Carson reservation, confirm this theory. (Courtesy of Cultural Resources.)

Hidden in the Dakota sandstone, a scientist discovered three-toed tracks tucked in the southern section of the reservation in 2009. This is one of 12 footprints that belong to a much larger trackway crossing the state between its northern and southern boundaries. Each footprint spans approximately seven inches in length, and scholars agree they belonged to a juvenile ornithopod dinosaur. These creatures dominated the North American landscape during the Cretaceous period. (Courtesy of James Kulbeth.)

The rugged rock formations also reveal evidence of prehistoric people who lived on the land before it became a military training ground. Rock art hidden mostly along the Turkey Creek canyon walls depicts geometric shapes, swirling designs, deer, and turkey. To paint the pictographs on rock faces, artists used a brownish-red color extracted from berries, and an earthy, yellow pigment obtained from ocher in the local sedimentary rock formations.

Other artists pecked their designs into the dark walls using two rocks together like a hammer and punch. To form just one symbol required hundreds of blows against the rock. This technique revealed the light-colored, newly exposed sandstone underneath, highlighting the rough designs classified as petroglyphs. The prehistoric pictures in the Turkey Creek Rock Art District date back 700 to 1,000 years.

Later, the plains Indians lived a nomadic life as their game determined their next campground. A teepee provided a portable home that could be put up and taken down with ease and carried to the next campsite. This image closely resembles what the eastern side of present-day Fort Carson looked like when Indians camped among the rolling grasslands. (Courtesy of William J. Carpenter.)

In a dry wash near a spring in the grassy terrain, a member of Carson's Cultural Resources discovered this amber and brown stone knife, in pristine condition, on the southeastern edge of the reservation. The jasper from which it was crafted comes from outside the Pikes Peak region. Hunters found the spring to be an ideal site as it attracted big game from the dry high plains. (Courtesy of James Kulbeth.)

This metate and mano, also found on the reservation, further confirms that Native Indian peoples once inhabited the land. For centuries, women used these stones as a tool to prepare flour. While holding the small mano, they pounded dry corn and other dried edibles against the heavy metate. Over repeated use, a depression formed in the middle of the base stone. (Angela Thaden Hahn.)

While native tribes traversed the area, Lt. Zebulon Pike attempted to summit the mountain that now bears his name. This illustration comes from a book of Pike's writings describing his adventurous 1806 expedition through the southwestern United States. Pike journaled his descent from the mountains back to his base camp as he and his men followed Turkey Creek and sheltered in the piñon-juniper woodlands among old Comanche campsites. (Courtesy of Mary Gay Humphreys.)

Half a century later, Henry Harkens homesteaded a parcel on present-day reservation land near the old Canon City and Turkey Creek Road. While he was constructing a sawmill, the infamous Espinoza brothers savagely murdered him. This 1863 gravestone, broken and repaired over time, marks Harkens's isolated burial place on a knoll overlooking a gulch now known as Dead Man's Canyon. (Courtesy of Charles Hatch.)

Within the next few decades, families trickled into the area and lived a hardscrabble existence among the dry, rocky, sandstone foothills on the west side of the present-day Army post. Settlers collected logs—the building material most readily available in their area—and crafted them into small homes like this Colorado cabin. Most of the time, these crude structures served as temporary shelters until stouter quarters could be built. (Courtesy of C.W. Talbot.)

The nearby Canon City and Turkey Creek wagon road became a stagecoach route in 1873 and carried passengers and light freight all the way to Denver. Today, Highway 115 follows much of this original course as it borders the west side of Fort Carson. Glendale, just outside the post's southwest corner, was the site of a stage stop along this route. (Courtesy of John C.H. Grabill.)

On the east side of what later became Fort Carson, Isaac and Edith Gaut lived on a ranch close to the town of Fountain. Ranches in these semi-arid meadows and rocky foothills proved suitable for grazing stock. The family's simple home appears to be partly constructed of adobe, a common early building material in the area. (Courtesy of the Fountain Valley Historical Society and Museum.)

At the Gaut ranch, a woman dressed warmly against the winter chill tends a large flock of chickens. No matter the weather, which could be very severe in these foothills next to the mountains, tending animals prevailed as a daily chore. The spacious coop allowed for fresh air ventilation by day and could be closed up against predators at night. (Courtesy of the Fountain Valley Historical Society and Museum.)

18

With additional land purchases, Fort Carson acquired the Turkey Creek Ranch, which included this beautiful bungalow built by Colorado Springs philanthropist Spencer Penrose, where his family had spent their weekends. The one-story structure embraces Spanish Revival architecture with a terra cotta tile roof and a stucco exterior painted white. In the early 1970s, soldiers of the Cavalry Horse Platoon used the dwelling as their bunkhouse. (Courtesy of James Drye.)

Spencer Penrose left his Philadelphia home with a desire to go west and soon amassed a fortune in gold and copper mining. He built several tourist attractions in the Colorado Springs area, such as the Cheyenne Mountain Zoo, the Pikes Peak Highway, and the Broadmoor Hotel. Additionally, Penrose established the El Pomar Foundation, which continues to offer grants to nonprofit organizations.

More purchases of land expanded Fort Carson's training range and included the former Mary Ellen and Mesa View Ranches. Eventually, the total acreage of the original reservation became 60,048 acres. As with the Turkey Creek Ranch structures, the Army made use of this log cabin, which became part of the Army's dog training school.

South of the Penrose place, local homesteaders gathered at Lytle on Turkey Creek, where a post office and schoolhouse served the community. After her graduation from high school, Coral Toothman (far left) taught at the Lytle School. Toothman resided in Fountain, about 10 miles northeast of Lytle as the crow flies, but most likely boarded with one of the ranch families in the Lytle community. (Courtesy of the Fountain Valley Historical Society and Museum.)

South of Lytle, Stone City thrived at the site of a stone quarry, much like the one pictured here, where miners cut Turkey Creek sandstone and shipped it by train to Pueblo. The white sandstone proved to be of high quality and became a popular construction material for strong edifices such as the third, and current, courthouse of Pueblo County, completed in 1912. (Courtesy of the US Geological Survey.)

The Stone City railroad station sat at the end of a spur off the Colorado-Kansas Railroad, which connected the nearby stone quarry to the city of Pueblo. For about 30 years, miners cut Dakota sandstone and collected valuable fire clay from the Turkey Creek mining district and sent it to construction sites as far away as Denver. The railroad to Stone City operated from 1909 until 1934. (Angela Thaden Hahn.)

These 1922 railroad maps point out the various historical locations in the area that later became part of Fort Carson's training range. The current installation occupies most of the area on the El Paso County map south of Colorado Springs, roughly bounded by Teller and Fremont Counties on the west and the Denver & Rio Grande (D&RG) Railroad on the east. The range extends a few miles south into Pueblo County as well. Dots mark Glendale, Fountain, Lytle, Stone City, and Kelker, which will be mentioned later. (Both, courtesy of George F. Cram.)

Two

TENACIOUS LIKE MISSOURI MULES

A military policeman (MP) stands guard outside the two-story Camp Carson headquarters. When construction of the Army post began in 1942, laborers completed this building first. They used heavy concrete blocks, and it was painted white. As the installation grew and continued to stay in operation, a newer and more modern redbrick headquarters building replaced this first structure.

Civilian laborers came from nearby cities and other states to build the original barracks and workspaces. They covered the buildings in wooden clapboard and painted them white. The workers completed the first sets of structures in July 1942, and the reactivated 89th Infantry Division moved in. The Army also hired civilians to work as guards and post exchange clerks. (Courtesy of Fountain Valley Historical Society and Museum.)

During the camp's construction, railroad workers laid a spur at the north end of the post near present-day B Street. The spur connected the installation to the Kelker siding on the D&RG Railroad. Ideally, if all equipment and supplies were ready, an Army brigade of 300 to 500 men could ship out for maneuvers in a day. (Angela Thaden Hahn.)

At the southern end of the post, the Camp Carson hospital sprawls over roughly 50 acres in 1943. Prussman Boulevard and present-day Mekong Street meet at right center. This cluster became the largest hospital complex in the United States during World War II. Nurses often used scooters and bicycles to traverse the long distances between wards. Miles of covered ramps and hallways connected the buildings together. Like the headquarters building, strong concrete construction blocks gave the hospital a life expectancy of 25 years. Indeed, it lasted well into the 1980s before it was replaced by the current Evans Army Community Hospital. Today, Building 6222 on Mekong Street is one of only five buildings remaining of the original bustling compound.

The 50th General Hospital on Camp Carson received an important visitor in 1943 when Pres. Franklin Roosevelt came to inspect its performance and operations. Accompanying the president were Brig. Gen. Thomas Finley in command of the 89th Infantry Division and Col. Wilfrid Blunt, post commander. Roosevelt fully supported the military and desired to increase its size from 137,000 troops in 1940 to 900,000.

Meanwhile, soldiers of the 89th Infantry Division trained on obstacle courses. A typical course contained items simulating the numerous objects a soldier might face in combat, such as bombed-out buildings, unstable rubble, and rough terrain. Trainees encountered obstacles to climb over, crawl under, balance on, hang from, and jump over. These exercises taught men how to navigate muddy water and precarious ropes and nets.

The natural terrain on Carson's reservation provided authentic obstacles, such as these fallen cottonwood trees. Downrange, soldiers experienced dry conditions with very little running water and extreme temperature fluctuations in accordance with the seasons. Trainees encountered sharp-pointed yucca and cactus on uneven ground sparsely pocked with rabbit and prairie dog holes. Meeting rattlesnakes, scorpions, and tarantulas was always a possibility. Soldiers learned how to cross over ground that rose up sharply to table-top formations and dropped suddenly into dry ravines.

Scaling a 15-foot embankment, men of the 89th Infantry Division reach out to their comrades, ready to lend a hand. Such commando exercises developed teamwork and problem-solving skills for young men who attempted to fill an essential role in a fighting unit. The 89th was the first major unit to be activated at Camp Carson. It was in fact a reactivated unit from World War I. Later, the 71st and 104th Infantry Divisions would train on these courses.

Soldiers of the 89th Infantry Division engineers at Camp Carson gather for training in 1943. Bridge-building has always been an important task for combat engineers. In order to form a crossing point for the rest of the division, these soldiers fasten together sections of a portable bridge, possibly dropped by aircraft.

Soldiers of the 89th Infantry Division artillery train downrange on a 75mm pack howitzer in the fall of 1943. This cannon, nicknamed "Little Dynamite," could fire five rounds per minute with a maximum range of about 5.5 miles. Considered a light gun, it was designed to be carried to a firing location in several parts and then reassembled for use.

Officers train in an experimental Ford amphibious jeep in Carson Lake. This vehicle was built in the early 1940s for versatility on land and in water. Soldiers referred to it as a "jeep in a bathtub," or a "seep." Although similar to the larger DUKW, or Duck amphibious trucks, personnel considered the seeps too slow and heavy on land. The production seep had become 900 pounds heavier than the original intended design. The seep also lacked sufficient boating abilities in open water because it could not handle more than slightly choppy waves. Therefore, the seep showed quite limited abilities in anticipated activities such as ferrying troops and cargo from ships offshore, then over a beach, and continuing inland. Consequently, the seep failed in field trials.

Cameraman Private Norbie of the 165th Signal Photo Detachment titled this photograph, "Want a Ride," with the following caption: "Yes but I will ride in the next water taxi, says newspaper correspondent. Col. Blunt, Post Commander, rides in the front seat with Lt. Roberts in the back seat." Norbie captured this image at Carson Lake on April 1, 1943, when the Army officers went on a test ride in the seep.

In October 1943, a cameraman snapped this image from a road named "Agony" by tank crews because the steep and winding passage became a long and agonizing climb while driving a heavy armored vehicle. This view looks north over the piedmont plains and tablelands that comprise most of the Army reservation, with Cheyenne Mountain on the left.

Carson's desert open range provided excellent ground for large weapons training. Surrounded by sharp-pointed yucca, these soldiers operate a light 37mm gun M3 during maneuvers as they practice guarding a road. They dug a depression in the dusty earth in which to steer the carriage and positioned the muzzle barely above the ground. A light gun could disable a tank by shooting it off its tracks. However, this gun proved to be ineffective against the heavy German Panzer tanks during World War II. As a result, this weapon was thoroughly obsolete by 1944 after just four years of standard use.

Early in the war, American artillerymen often trained with field pieces left over from World War I. Hidden beneath staked camouflage netting, men from the 89th Division load a round into the breach of a vintage 155mm howitzer. Compared with the lighter guns, this heavy artillery piece had a greater firing range, with each round weighing up to 95 pounds. The 155 was well-suited for destroying protected enemy targets and striking rear areas. Newer guns were introduced before the unit saw combat in Europe; nevertheless, this gun remains a mainstay of heavy artillery to this day.

Jeeps served a variety of roles during World War II. Above, medics find the jeep useful as an ambulance under emergency conditions by accommodating four litters at one time. Along with infantrymen, hospital personnel also received training to operate under battlefield conditions. Personnel needed to be able to set up and take down a field hospital so that it could be moved to wherever the troops were. Below, members of the 50th General Hospital unload a patient from a Dodge ambulance during a field exercise. The Dodge Power Wagon became a reliable three-quarter-ton truck and was used for a variety of purposes.

Soldiers in training learned how to survive being overrun by armored vehicles. In April 1943, an infantryman emerges from his foxhole in the middle of the road to peek at the M10 tank roaring toward him. After it passes, the soldier hurls a training grenade at the vulnerable back end of the tank. Upon impact, the outer glass layer of a round grenade, known as a sticky bomb, shattered and exposed an adhesive coating, which caused the grenade to stick to the surface upon which it landed. These grenades were effective against small armored vehicles and the tracks of larger tanks.

A tank crew practices firing downrange using their fingers as their only ear protection. Many soldiers during this era suffered hearing loss. Indeed, noise-induced hearing loss and tinnitus were some of the most common service-related disabilities. Almost every soldier was exposed to hazardous noise levels at some point in their career. Ironically, hearing loss significantly impaired a soldier's effectiveness on the battlefield.

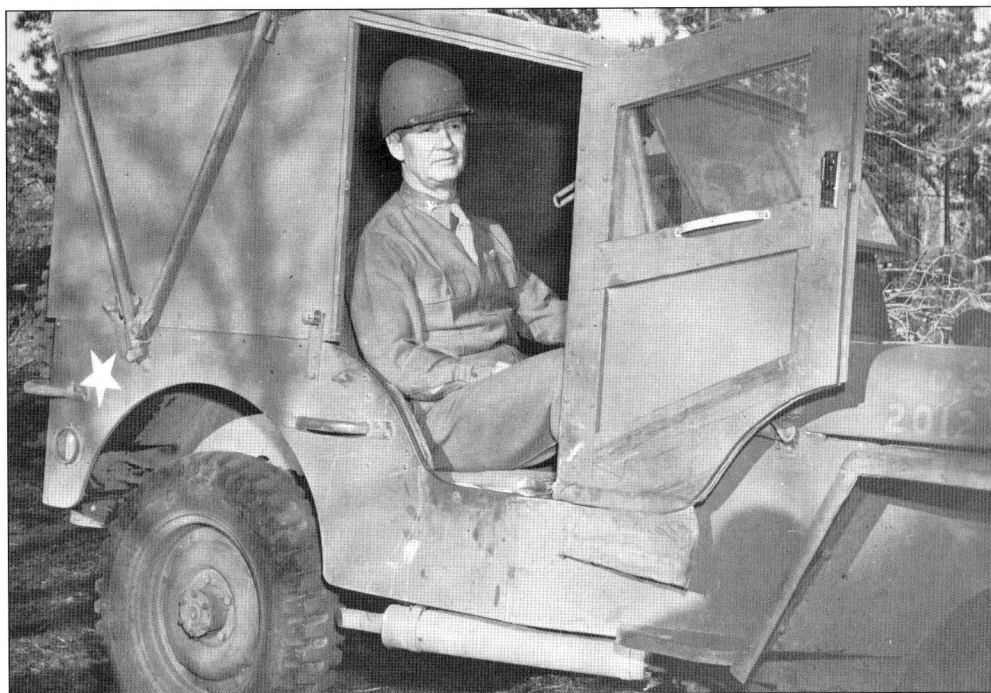

In April 1943, Col. C.L. Irwin of Laramie, Wyoming, arrives at bivouac by jeep. Irwin was shortly thereafter promoted to brigadier general and assistant division commander of the 89th Division. Bivouac was a common word to describe a temporary military encampment under little or no shelter and was very similar to long-term, primitive camping downrange away from the barracks.

The unidentified signal corps cameraman who captured this image titled it "Buddies": "Where there are officers and men there is at least one dog around. Lt. Col. William C. Bliss, Ordinance officer, has found a buddy on the 89th Division Headquarters Co. bivouac. April 1943." Stray dogs often found their way to an Army post. Most soldiers could not resist the company of the four-legged allies and readily adopted them as part of the camp family.

A tank commander calls in for coordinates and his crew waits for a signal as they engage in wartime exercises downrange on Camp Carson. They have bestowed the name of "Big Boy" to the 105mm howitzer motor carriage M7 Priest, of which they have been put in charge. A large tarp covers the ground where members of the crew can work with ammunition, keeping it fairly free of debris. Each shell weighed about 25 pounds.

A soldier trains at the firing range with an M1903 Springfield rifle. This bolt-action firearm became the standard issue rifle during the First World War and was officially replaced in 1937 by the M1. However, by the time the United States entered World War II, there were not enough M1 rifles to arm all troops. As a result, the M1903 Springfield remained in service.

Nurses of the Women's Army Corps (WAC) parade past the hospital at Camp Carson. WACs arrived at the camp before any other soldiers. Political and military leaders realized that women could supply additional resources while so many men served outside the country in fighting the war on two fronts. The WACs became an essential part of operations and eventually merged with the general population of soldiers in 1979.

A soldier from the 89th Division trains on an M18 recoilless rifle. The design of this weapon allows the operator to fire 57mm projectiles from the shoulder or from a prone position. However, the M1917A1 machine gun mount shown in this image kept the gun most stable. The instructor guides the trainee's head as he learns how to line up the sights of the weapon.

A squad of men sits on the ground while listening to firearm instruction from their platoon sergeant. The distinctive cover on their heads identifies their olive drab, military 1942 fatigue uniform. This new design featured large cargo pockets on the jacket and trousers. The fabric was woven into a rip-resistant, herringbone twill pattern that held up well in rough field training conditions. Eyelets on both sides of the hat provided ventilation.

Foreign allied military units often come to the United States for training in advanced tactics and strategies that they take back home to teach to their own armies. This helps them to be better able to fight their own wars without an American presence. In about 1943, Camp Carson hosted the 122nd Infantry Greek Battalion. In an outdoor ceremony on Carson's parade field, the battalion became an official unit and received its colors. The battalion commander gave a speech, and at its conclusion, the battalion marched past the reviewing stand. With fists held high, the Greek soldiers vowed to use their newly acquired military skills to avenge the invasion of their native land by fascist Italy and Nazi Germany.

A unit of the 89th Division engages in field day activities in October 1943. Men came together on Field Day to compete in the skills they had recently learned during their basic training. The exercise shown here simulates the beams that may be all that is left of a bombed bridge. Trainees learned how to get themselves across the river to the other side without falling into the water below.

Army basic training included forced marches. A unit of the 89th Division participates in a forced march in November 1943. Such marches simulated the need for a unit to relocate quickly. Therefore, on these march exercises, the men carried their full gear for distances of 10 miles or more and often at a quick pace. They kept a certain distance apart to reduce casualties if the enemy attacked.

A complete basic training course consisted of 17 weeks and included a variety of infantry exercises. In this image, the 35th Infantry of the 89th Division participates in an infiltration course in February 1943. Typically, soldiers crawled on the ground while all around them, small dynamite charges simulated the artillery explosions in realistic battle conditions.

Members of the 89th Division mobilize their equipment with M3A4 hand carts in October 1943. The carts were made of an aluminum alloy and were mounted on two sets of motorcycle wheels and tires. The carts in this image carry rifles, a shovel, and bundles covered in tarps. Two carts could fit in the back of a two-and-a-half-ton truck on the floor between the two side benches of troops. When trucks could no longer proceed, the men loaded the carts with gear and transported them by manpower.

In the summer of 1943, a soldier engages in survival training as he uses a long pole to glide through the water on a raft made of canvass tarpaulins lashed to a crude frame of brushwood. The small spring-fed lake is one of a few located in the southern half of the Carson reservation.

A medical command unit of the 89th Infantry Division at Camp Carson holds formation at a field sanitation training site set up for trainees. Various types of incinerators, latrines, water treatment systems, and food waste disposal units are assembled to teach soldiers how to stay healthy in bivouac conditions, depending on their length of stay at one encampment.

A muddy motor pool holds a fleet of parked GMC CCKW two-and-a-half-ton six-by-six cargo trucks. Known as a deuce-and-a-half, these heavy-duty vehicles conquered off-road conditions. The Army had large numbers of CCKW trucks built from 1941 to 1945 and used them to haul cargo and troops during World War II and the Korean War. They remained in service until the 1960s and proved to be useful trucks. Eventually, the M35 series of two-and-a-half-ton trucks replaced the CCKW model.

In April 1943, a tank destroyer convoy of the 89th Division rumbles south down a dusty road alongside Carson's motor pools. Tank training involved shooting projectiles downrange, but within the cantonment, tank crews learned how to convoy and load the tracked vehicles on flatbed train cars.

Soldiers ride in a three-quarter-ton WC51 (weapons carrier) truck with a 37mm gun in tow as they head for the firing range on Fort Carson in April 1943. This four-by-four truck improved upon the half-ton truck used in World War I. The WC51 body came with seats so that soldiers could be carried as well as supplies. Personnel used the three-quarter-ton truck heavily during World War II. The division sergeant major resided in the home in the background on the left. The foreman of the former Cheyenne Valley Ranch once lived there.

Trainees in the 3rd Battalion of the 353rd Infantry, 89th Division, participate in an advance march on a mock foreign village at Camp Carson in April 1943. They chose a dry arroyo bed, surrounded by high cliffs, for concealment. A combat unit uses this type of march when they want to quickly make direct contact with the enemy.

Units of the 89th Division storm the mock French village at Camp Carson in April 1943. These roughly constructed buildings simulated a European village, where street fighting and house to house combat were common. In this image, soldiers practice overcoming a barbed wire obstacle by crawling underneath without getting snagged.

In the summer of 1943, Camp Carson's mock French village included a cemetery and chapel. Beauclaire became the name of the make-believe hamlet, in memory of the French village taken by the former 89th Division in World War I. After World War II, the producers of the movie One Minute to Zero transformed the site into a Korean village for the movie set. The Asian village then became a combat training ground for Korea-bound troops.

As soldiers dodge sniper fire from the chapel steeple, they decide not to crawl under the barbed wire but instead charge over it in an attempt to take the Nazi position in a combat rehearsal. Barbed wire had played an essential role in World War I, and it continued to be used in World War II. After conquering the barbed wire, soldiers were trained to clear the simulated village house by house. Some soldiers say that this is the most dreaded kind of fighting. While the troops engaged in mock combat fighting, unit leaders watched from bleachers where they could observe and give instruction when needed.

At Beauclaire in April 1943, soldiers were trained to enter buildings and confront whoever may be on the other side—friend or foe. Unit leaders rigged plywood cutouts of enemy combatants and town civilians on ropes and pulleys. These were raised without warning to give the soldiers practice in identifying the enemy and in reacting accordingly.

Once a structure was seized, it had to be defended from enemy repossession. Units of the 89th Division hold a building they have just captured from "the Nazis" in Beauclaire during Carson war games in the summer of 1943.

At Beauclaire in April 1943, members of the 3rd Battalion, 353rd Infantry, 89th Division, evade fake land mines as the smoke of battle drifts over the village. Soldiers run from house to house, seeking the enemy, as they take possession of the town and charge the Gestapo headquarters. Trainees can be seen inside the windows of the buildings. The battle was won when the Swastika over the courthouse, at center below, came down. The mock village included details such as picket fences, outhouses, street signs, and juniper trees planted every few feet along the main street. These realistic training activities were essential in preparing the troops for battle in Europe.

Four medics with the 89th Infantry Division train at Beauclaire in 1943. In this image, they practice using a litter to evacuate a wounded soldier from the ongoing battle while ducking enemy fire. They likely fastened the stretcher to a jeep for transport to a nearby field hospital.

Maneuvers afforded time off from training. Spending as much as several weeks downrange, GIs appreciated diversions delivered by the mobile post exchange. This convenience allowed the men to purchase cigarettes, sweets, and useful items like pens and writing pads while still in bivouac.

The Camp Carson gymnasium is one of the oldest structures remaining on the post, where indoor sports such as basketball and wrestling were offered. It is located on Hogan Street between Specker and Wetzel Avenues. Now known as the Bill Reed Special Events Center (SEV), this facility serves as a gathering place where soldiers returning from combat reunite with their waiting families.

Two soldiers head southeast toward H Street after having crossed one of the former ranch irrigation ditches that extend through the post. Construction continues on the right. On this crisp winter day, the bright blue sky holds some clouds, but the temperature is too frigid to melt the remains of snow pushed to the sides of the street. The region's mild winter climate made this an attractive spot for a year-round military training base.

The large "PW" on the back of these men's jackets identifies them as prisoners of war. Camp Carson maintained a POW camp for German, Italian, and Japanese prisoners during World War II, beginning in 1943. They stayed in barracks built at the northwest corner of the post, bordered by Highway 115 and present-day Academy Boulevard. The US Army provided work for the prisoners to perform in the camp bakery and the laundry facilities. They built furniture in a prison woodworking shop and published a magazine in the German language. Interestingly, the prisoners of war quartered at Camp Carson alleviated the labor shortage caused by the entry of so many of the state's young men into military service. The prisoners engaged in general farm work around the region and earned about 80¢ each day. The Venetucci and Wilson farms and other local farmers in the neighboring Fountain Valley employed the prisoners to clear fields for planting, labor in beet fields, and harvest corn. After the war, many of the prisoners had favorable things to say about their imprisonment at Camp Carson.

Camp Carson boasted eight chapels to provide for the religious needs of its thousands of soldiers. The typical World War II–era chapel resembled the small country churches scattered all across America's countryside in places that the troops called home. Catholic, Protestant, and Jewish chaplains were available to personnel belonging to each respective religion.

Fellow officers form the traditional saber arch at the exit to one of the post chapels for this newly wedded couple. Soldiers soon to depart for overseas service often rushed to marry before shipping out. The increase in wartime and postwar weddings resulted in the US 1950s baby-boom generation.

In December 1942, men in military dress uniforms wait to board a white double-decker bus transformed from a tractor and semi-trailer rig. It held up to 260 soldiers and transported them to various parts of the camp reservation, and to central points where they could transfer to city busses going to Colorado Springs and Fountain. The government issue bus contained 700 square feet of floor space, 100 seats, and room to stand if all the seats were full. The interior view shows where the seats were installed and how lumber was used to crudely construct stairs, railings, and barriers within the trailer. The GIs referred to these busses as cattle cars and broke into sounds of mooing cows while riding inside.

At left, a soldier prepares to ride into Colorado Springs with a pass from his unit. Troops boarded busses that departed regularly from the north gate of the post. MPs were stationed in town to check paperwork and to help the GIs get back on the busses in time to arrive at the camp before taps. Below, 1st Sgt. Zerrugi pauses at a lamppost on Pikes Peak Avenue in the summer of 1943 while other soldiers in uniform and their dates gather outside the Balcom Café. This spot is east of Tejon Street. To the west, the two spires of the Antlers Hotel peek up in the background. The romantic comedy *The Youngest Profession*, starring Agnes Moorehead, among others, was showing at the Ute Theater. Currently, US Bank occupies the tall building at far left.

Camp Carson's mules first arrived from Nebraska in July 1942 when construction of the post was completed. Above, C Company demonstrates how to pack a 75mm howitzer that has been broken down into six mule loads for transport in mountainous terrain. Early in the war, Army soldiers still wore World War I–style helmets. The common M1 steel pot helmet had not yet been issued at Camp Carson. The packers became quite attached to their mules, and this adoration spilled over into the local community as well. Army mules became a popular parade attraction in Colorado Springs. Below, a mule train marches in the 1942 Veterans Day parade. They carry a 75mm pack howitzer south on Cascade Avenue past the famed Antlers Hotel. In a show of respect, the men look toward the dignitaries on the reviewing stand. Included in the front rank on the stand are Major General Gill, commander of the 89th Division; General Brittingham, the division artillery commander; and Brigadier General Finley, assistant to the major general.

Camp Carson's most famous Army mule, Hambone, stands with his handler, Master Sgt. Lawrence "Pop" Cnossen. Hambone served for 13 years at Camp Carson in the 4th Field Artillery Pack Battalion. He was born in Missouri in 1932 and was known for his beautiful silvery-white coat and for his talent as a skilled jumper.

Hambone's white coat easily identifies him among his fellow pack mules. He often carried the first sergeant on treks up Ute Pass to Camp Hale. This provided excellent cross-country, long-distance, endurance training for the animals. The four-legged soldiers proved to be a valuable part of the military because they could make their way through terrain where vehicles could not pass.

Men from the 604th and 605th Field Artillery (Pack) learned how to break wild mules and care for them at Camp Carson from 1942 to 1956. Mules were trained to carry a field pack long distances over rough terrain. In addition to transporting equipment, weapons, and supplies for the men, they also carried their own feed. Farm boys with experience in stock animals were especially valued for this military occupational specialty. They learned how to tie specific knots used in attaching the packs to the mules. Training spanned six to eight weeks and started with introducing the animals to the feel of the rigging. Later, heavier packs were placed on their backs until they became accustomed to the load. As part of their training exercises, Carson pack mule trains of up to 300 animals climbed Barr Trail to the 14,000-foot summit of Pikes Peak. Someone humorously wrote long ago that Carson's mules were paid $40 per day, while their handlers received only $2 for the same amount of time.

Mules and men representing the US Army in World War II take their place in formation as they prepare to enter the Carson parade field. In front of them are men representing World War I soldiers with horse-drawn weaponry, servicemen of the Spanish-American War, and one lone horseman dressed in an 1870s cavalry uniform. Leading the parade is the Camp Carson Mounted

Color Guard. These groups of men help to portray the history of the Army. This formation would have been an appropriate display during an Armed Forces Day celebration when civilians were traditionally invited to come learn about their Army.

During the 1940s and 1950s, patriotism ran high for the country and its servicemen and women. The city of Colorado Springs hosted a few parades each year on patriotic holidays. They were well-attended by local civilians. Here, deuce-and-a-half trucks carrying soldiers and hauling 105mm howitzers drive across Pikes Peak Avenue. Note the ticket booth for the Pikes Peak Cog Railway and the Cheyenne Mountain Highway by the Winthrop shoe store.

Hambone jumps a hurdle at one of his competitions. He required a more dignified name for these events and entered as Hamilton T. Bone. Legend has it that on at least one occasion, he had to return his championship ribbons after being disqualified for not being a horse. He gained national notoriety in 1949 when *Life* magazine featured him in an article.

Upon his arrival at Camp Carson in April 1952, Gen. Mark W. Clark, chief of Army field forces, inspects the honor guard appointed to receive him. Guard commander Capt. Paul M. Ireland Jr. accompanies General Clark down the ranks. General Clark was named the chief in 1949 after impressive service during World War II. His Army career was not without controversy, but he was much admired by General Eisenhower.

When World War II ended, Camp Carson was considered for closure; however, talks between the city fathers and the War Department resulted in keeping the camp active. In April 1953, civilian and military dignitaries gather outside the headquarters building. From left to right are Verne Johnson, ? Willie, Al Dalpiaz, Vernon Hollenbeck, Robert Newman, John M. Biery, Eugene Martin, Joseph Reich, and General Keyser.

The 4th Field Artillery Pack Battalion arrived at Camp Carson in 1947 after serving in the Philippines and the Panama Canal Zone. At this new duty station, they supported winter maneuvers at Camp Hale near Leadville, Colorado. In May 1943, members of the 4th Field Artillery Pack Battalion firing battery load a 75mm pack howitzer tube on the back of a mule in preparation for the Armed Forces Day parade in Colorado Springs. The barrel was carried using special tack and equipment.

Three

VALUABLE LIKE
RED DIAMONDS

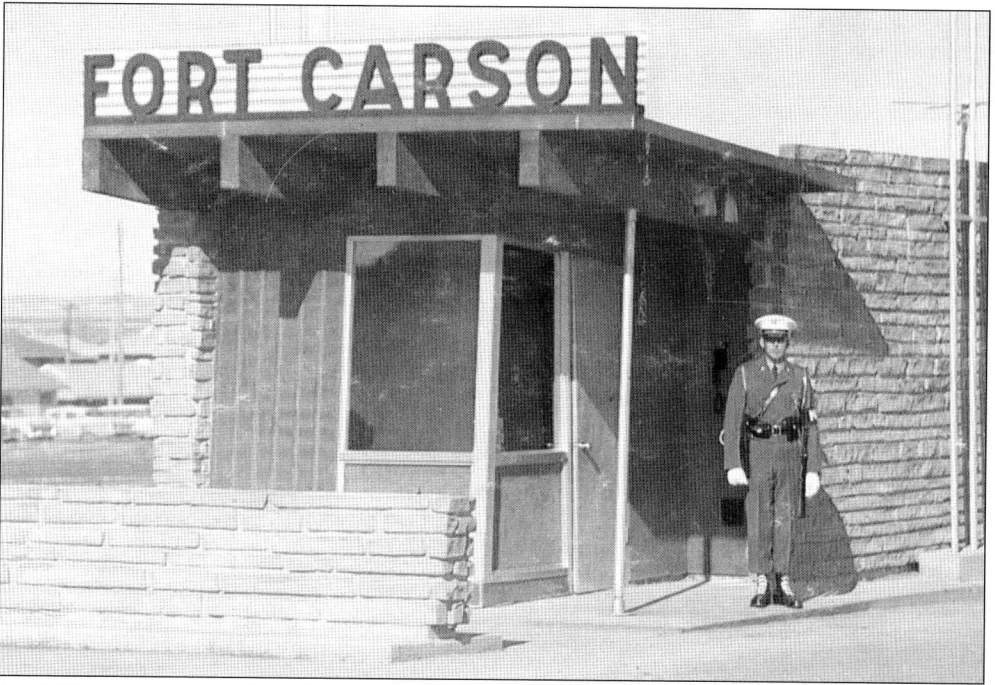

An MP, wearing a uniform from the 1950s, stands guard outside one of Fort Carson's gates. The Korean War ended with the country in a recession. Budget cuts in the defense department threatened the demise of Camp Carson, but in 1954, Congress decided to keep the military post and reclassify it as a fort. Funds were appropriated to replace the temporary camp barracks with brick buildings and construct family housing.

Hundreds of local residents and active duty servicemen turn out for a parade in 1956, undeterred by humming construction projects in downtown Colorado Springs. With the reviewing stand perched behind them in the middle of Pikes Peak Avenue, the pack mule unit heads south on Tejon Street next to the Exchange National Bank. *The Birds and the Bees* highlights the marquee at the theater in the background, which shares street space with jewelry stores and the *Gazette Telegraph* newspaper office.

Hambone and his handler Master Sergeant Cnossen prepare for Hambone's retirement ceremony in February 1957, when the Army phased out mules and replaced them with helicopters. Carson mule handlers loved Hambone for his intelligence, loyalty, and hard work. However, they also noted that he was after all a mule and could be somewhat ornery.

When Army pack mules became less efficient than mechanized vehicles, Carson's mule battalion was retired. Above, Hambone takes some time to inspect his fellow artillery pack mules at the retirement ceremony. As he strolls by, his comrades stand at attention, with ears pointed forward as they pay respect to their honored friend. The ceremony took place at the mule barns on the northeast portion of Fort Carson near the railroad depot. Below, Hambone's handler allows him to spend some last moments with his companions, some of whom he had worked with for many years. The local Al Kaly Shriners bought most of the animals to start their first mule train. The Shriners' mules are now stabled at the large red and white dairy barn just south of Colorado Springs and visible from Interstate 25. The Carson barns were razed soon after the mules departed. Sometimes these mules returned to the post for ceremonial events and headed directly toward the familiar surroundings of their former stalls.

Above, Trotter, named for his ability to trot long distances, stands next to Hambone as they both retired on the same day. Trotter went on to become the mascot for the US Military Academy at West Point and served in that capacity until 1972. Below, a tearful Master Sergeant Cnossen hands the reins of his good friend Hambone over to W. Thayer Tutt, president of the Broadmoor Hotel, as he accepted the famed mule from Maj. Gen. H.P. Storke, commanding general of Fort Carson and the 9th Infantry Division.

Shortly after his Army retirement, Hambone arrived at his new home at the Broadmoor Stables. The famous Broadmoor Hotel appears in the background of this image. Shown holding the reins is J.D. Ackerman, president of the Exchange National Bank in Colorado Springs, and a member of the Rodeo Association. Hambone went on to become the star attraction at the Pikes Peak or Bust Rodeo and also participated in events during Frontier Days in Cheyenne, Wyoming.

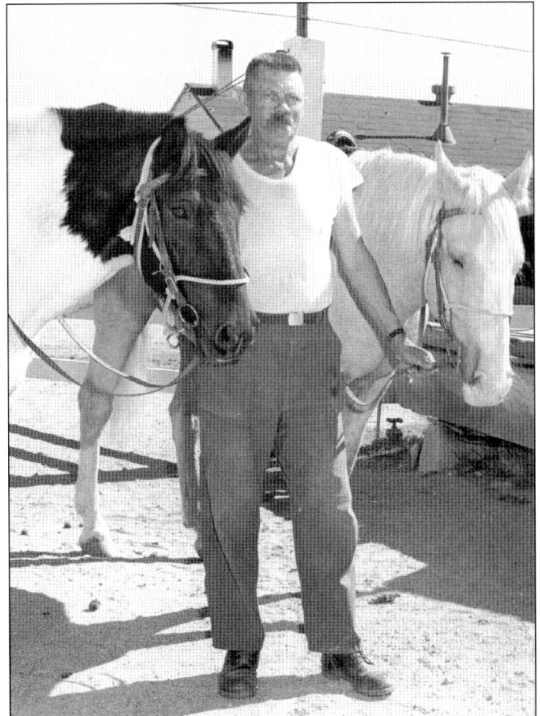

After the departure of Hambone and the pack mules, retired Master Sergeant Cnossen became the stable sergeant at the Fort Carson Riding Academy. He is pictured here in October 1961 with two of his new equine friends. He lived at the stables and took his meals there while employed by the Army as a civilian. He was at home among the animals, and it was said that for four years, Cnossen never left the ranch.

The only Army dog training center in the United States was created in 1951 and was located at the old Mary Ellen Ranch within the borders of Fort Carson. Training lasted from eight to twelve weeks, where canine scouts learned to use their sense of smell to detect the enemy. Messenger dogs learned to carry written communications, food, ammunition, and medical supplies. Sentry dogs were trained to patrol warehouses and ammo dumps and attack any intruders. The center sent 30 of its graduates every year to Army units. The Air Force received an annual amount of 600 sentry dogs that had been trained at the center. Just before closing in 1957, the center had 25 scout dog platoons. In that year, dog training for the military was transferred to the Air Force, and Fort Carson's dog training center closed.

The 9th Infantry Division color guard and flag team stand in formation outside Carson's headquarters building. The division insignia attached to the headquarters' outer wall indicates its current occupants. The 9th Infantry was nicknamed the "Old Reliables" for its strong and unyielding service in several battle engagements during World War II. The 9th called Fort Carson its home in the early 1960s.

Members of the 9th Division train on the M20 Super Bazooka. The instructor watches the two-man operation as the soldier on the right loads a 3.5-inch missile in the back end of the portable anti-tank weapon resting on the shoulder of the triggerman, who is sighting his target. The trainees wear masks to protect against any mishaps while learning to master the weapon.

New recruits arrive by bus for training at Fort Carson in the 1960s. Carefully selected and well-qualified instructors conducted all training at Fort Carson. Most of them were combat veterans of the Korean War, and many of them saw action in World War II. In addition to combat training on equipment, great emphasis was placed on the development of esprit de corps for the purpose of producing in the trainee a sense of identification with his unit and a sense of belonging.

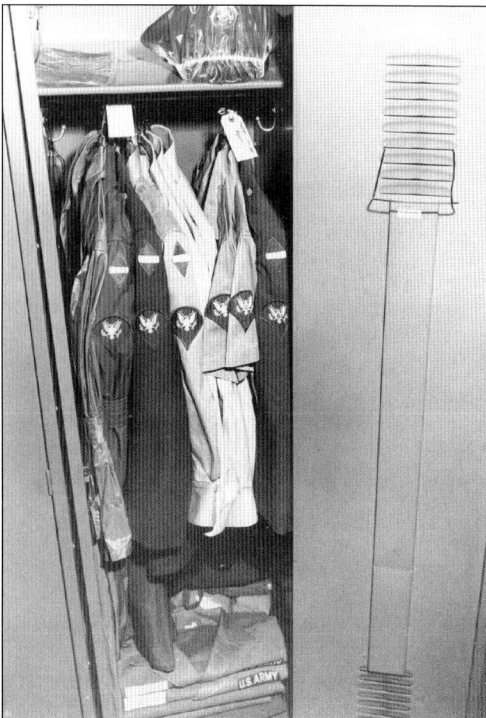

A peek inside this locker shows the uniforms of a soldier assigned to the 5th Infantry Division, evidenced by the red diamond patch on the upper sleeve. The Red Diamond Division was stationed at Fort Carson from 1962 to 1970. The triangular-shaped patch on the sleeve indicates this soldier held the rank of Specialist 4. While his dress shirts and jackets hang from the rod, his starched and ironed work fatigues lay neatly folded on the locker floor.

After serving as commanding general of 1st Corps in Korea and a brief assignment at the Pentagon, Maj. Gen. John A. Heintges assumed command of Fort Carson's 5th Infantry Division (Mechanized) in 1963. He expressed that he was proud and deeply honored to take command of such a famous division at such a historic post as Fort Carson, noting that the post had always been known as ideal for training. At the time of his death in 1993, he was a resident of nearby Colorado Springs.

MAJOR GENERAL JOHN A. HEINTGES
Commanding General
5th Infantry Division (Mechanized)

Two years after the Red Devils (5th Infantry Division) arrived at Fort Carson, General Heintges spoke to the 10,000 soldiers who made up the division at an anniversary parade in Pershing Field. Heintges commended the division for its role in the success of the reorganization of the 5th Division under the modern Reorganization Objective Army Division concept. This new program organized divisions for flexible response to handle battle situations, including nuclear war.

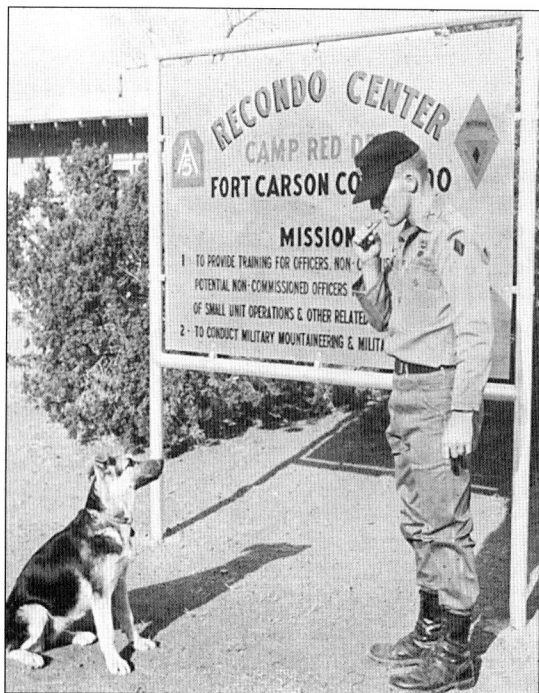

Camp Red Devil opened in 1966 under the direction of the 5th Division for the purpose of providing more year-round training for soldiers in a field environment. As the Army again struggled under looming budget cuts, the evaluation board came to Colorado Springs and stayed at the Broadmoor Hotel, where they found the area to be quite pleasant. With the growing need of more combat troops in Southeast Asia, they decided to keep the post open, and in 1962 the 5th Infantry was reactivated at Fort Carson and became the Army's first mechanized division.

Red diamonds painted on the tanks and jeep in this image identify them as 5th Infantry vehicles. In June 1969, formations of sparkling clean armored vehicles drive the dirt track circling Pershing Field. The full-scale division parade included formations of men and their equipment in a show of military might and unity.

Above, the rear side of the Soldiers Memorial Chapel, built in 1967, rises above a drainage ditch. It is located on the corner of Nelson Boulevard and Martinez Street, not far from Gate 1. Today, it offers both Catholic and Protestant services to personnel and their families. Often, the sanctuary is chosen to hold many of the funerals for fallen soldiers. This chapel and other Army chapels of the same design are known for their beautifully designed stained-glass windows. Protective layers of plexiglass were installed to guard against the region's infamous hailstorms.

Fort Carson soldiers stack arms while downrange. There are times when they must separate from their weapons, but instead of laying them on the ground, they stack them like a tripod into groups of three. The M1 Garand rifle, and many models that came before it, had a stacking swivel on the muzzle. This C-shaped ring was used to attach all three rifles together at the muzzle.

Search-and-seizure operations are carried out at this mock church as Spec. Dallas Kin of the 5th Administration Company prepares to enter the structure made of mud and thatch. James M. Isley, 4th Battalion, 84th Artillery, covers his comrade. In the 1960s, soldiers at Fort Carson trained at the mock village downrange. It had been reconstructed into a simulated Vietnamese village and renamed Bung Cong. Booby traps and land mines were added to the layout, and armored personnel carriers and helicopters participated in the mock assault.

Within the cantonment area of Fort Carson, artillerymen gather on a chilly winter morning in their winter parkas, M1951 pile hats, and black leather thermal gloves. The soldier on the left wears a 5th Army patch. He and two others discuss various metal components associated with the M101A1 105 mm howitzer behind them. This cannon was a new redesign of the M2A1 gun.

Soldiers train on the M20 Super Bazooka at the Carson firing range. Instructors stand beside each trainee to guide in proper foot positions while firing the weapon from the shoulder. The gun weighed 12 pounds and was an improvement of the earlier model that proved ineffective against Soviet tanks during the Korean War.

A keen eye and a magnifying glass help the Fort Carson enthusiast spot familiar places in this panoramic view of the cantonment from the 1960s. At the bottom, Yano Street runs from left to right, crossing Magrath Avenue on the left and intersecting Minnick Street on the right, with the motor pools sandwiched between these two streets all the way up. Parallel to and left of Magrath is Barkley Avenue. Stacked between Magrath and Barkley are the original clapboard barracks and workspaces of the 1940s. Those structures have since been replaced by barracks made of brick. Along the left side of Barkley runs Clover Ditch, an irrigation canal left over from former ranches. Left of the ditch is Pershing Field, bounded on the other side by Specker Avenue. Also visible are one of the original chapels on Barkley, a later chapel at the corner of Ellis Street and O'Connell Boulevard, the train depot and warehouses at the north end of the post, the indoor swimming pool on Nelson Boulevard, and the noncommissioned officers quarters just south of Gate 2, Highway 115, and the Broadmoor Hotel.

A Carson soldier engages in cold-weather combat training with an obstacle of barbed wire before him. He wears a hooded parka-type overcoat and an M1 helmet without any camouflage. He carries an M1 rifle and no gear on his back. Behind him could be smoke and dust from mock enemy fire. Or the soldier could be running from a chemical attack, and his free-hanging sack may contain a gas mask.

These soldiers are wearing the helmets of the military police corps. MPs during this era were particularly supportive during the Tet Offensive in Vietnam. Within the United States, the MPs served as the uniformed law enforcement branch of the Army, performing the same duties on post as civilian police units. They also provided escorts for convoys and personnel.

Pershing field was large enough for the whole division to turn out in formation at special ceremonies such as a change of command. The dirt track allowed for vehicles to pass in review of commanding officers, distinguished guests, and other spectators sitting in bleachers. Pershing Field was once much larger than it is today. The section pictured here is bounded by Specker Avenue, Ellis Street, Barkley Avenue, and O'Connell Boulevard and is no longer an open field. The three-story buildings behind the troop formations are still there today. O'Connell Boulevard can be seen on the left between the buildings and the mountain, heading west toward Gate 2 and Highway 115.

The red diamond, seen prominently from above, marks the headquarters building as the home of the 5th Infantry Division. After entering the post on Nelson Boulevard at Gate 1, Division Headquarters was found at the east end of the road, where it became the building's driveway. Flags from each state in the Union line three sides of the courtyard.

The 5th Division Band stands at parade rest outside the division headquarters building as they await an important visitor to the Army post. The 5th Army and 5th Division insignia are attached to the outer wall during their residency at the post. Band members wear their khaki summer garrison duty uniform.

Four

CHARMING LIKE HOME, SWEET HOME

In 1970, the 5th Infantry Division moved on, and the 4th Infantry Division took its place at Fort Carson. Shortly thereafter, a bronze statue depicting a soldier dressed in tropical combat uniform, commonly known as jungle fatigues, was added to the headquarters courtyard to honor the noncommissioned officers of the 4th and their recent participation in the Vietnam War.

Quarters 5512 D is part of a building that housed four families and is located at the corner of Harr Avenue and Aachen Drive in the noncommissioned officers' (NCO) family housing area near Gate 2. Author Angela Thaden Hahn moved with her family to this small, three-bedroom apartment in 1970. She and her brother walked across the street to the elementary school and explored the hills in the back. (Courtesy of Nancy Thaden.)

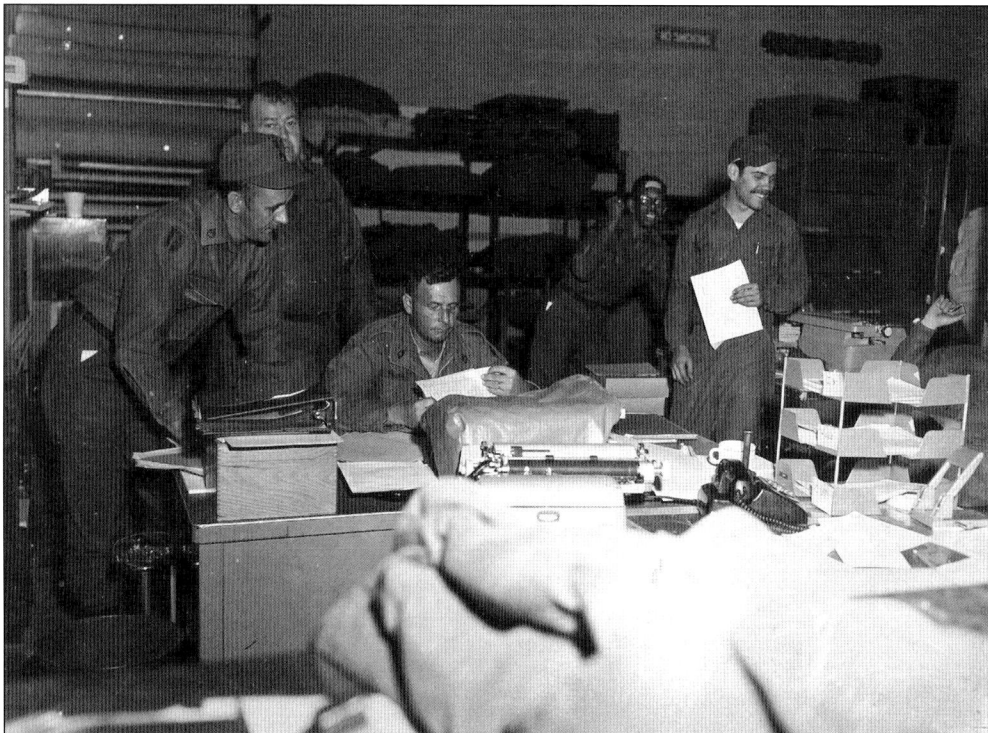

Sgt. 1st Class Les Thaden and a fellow soldier look over a unit's Table of Organization and Equipment as they prepare to conduct an inspection to confirm a unit's combat readiness. These men were members of the Inspector General Team. Sergeant Thaden brought to the team his 20 years of experience in a supply MOS. (Courtesy of Nancy Thaden.)

Uniformed soldiers returning from the Vietnam War were a common sight at airports across the country in the early 1970s. They arrived by commercial flights, sometimes alone and sometimes in small groups. Happy to be back home, they often stooped down to kiss the ground once their feet hit the tarmac. They were greeted by families whose usual communication for the past year had been by letters posted by airmail.

West Point graduate Bernard W. Rogers served as commanding general of Fort Carson in the early 1970s. While at Carson, he developed strategies to help the Army adjust to an all-volunteer force for a new generation of soldiers. While he commanded the 4th Infantry, he eliminated routines that he believed served no useful purpose, such as reveille, roll calls, and Saturday inspections.

Ann Ellen Rogers, the wife of commanding general Rogers, receives a beaded medallion from a representative of a native Indian tribe. In the early 1970s, Fort Carson hosted the annual Pikes Peak Intertribal Homecoming Encampment at Old Reliable Park. The four-day event followed the old Indian powwow tradition of tribes gathering for social purposes such as dancing and games. The modern powwow was open to the public and included dance contests for all ages, arts and crafts booths, some of the country's leading Indian dignitaries, and the Miss Indian Pikes Peak contest. In 1972, the vice-chairman of the Intertribal Club said, "The aim [of the event] is to educate Whites to the Indian culture and ways, and also to expose the Indians to the White progressive ways. It is also a way to let them see each other at their best." Thousands of people attended this event. The 1972 powwow hosted Miss Indian America Norma Begay from Wyoming and Chief Leo Vincenti of the Jicarilla Apaches from New Mexico.

Hambone died at Turkey Creek Ranch in March 1971 at the age of 38. He was mourned by many and was buried with full military honors near the 4th Division Artillery headquarters on Magrath Avenue. The 52nd Engineer Battalion constructed his memorial, pictured above, of stone quarried on the reservation, and dedicated it in May. Hambone had returned to Fort Carson in 1970 to retire from civilian life, where he lived out the rest of his days. At right, Hambone's fellow soldier, Wind River, was one of the last two remaining Army mules. He had served in the 35th Quartermaster (Pack) and retired with the rest of Fort Carson's pack mules in 1957 when his detachment was deactivated. Wind River died in November 1978.

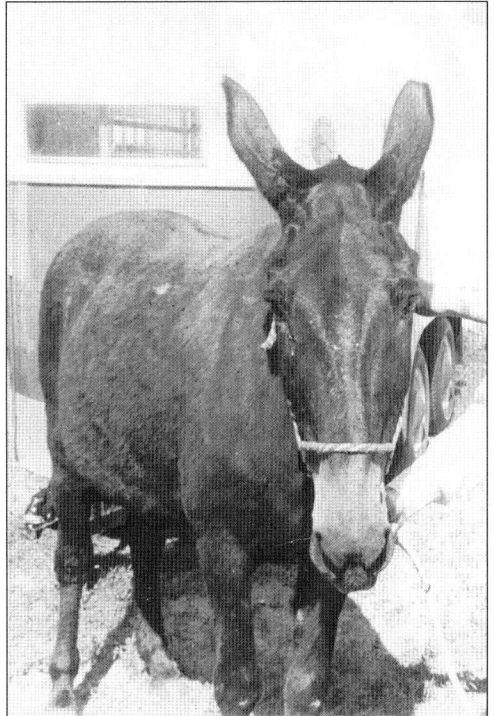

Above, members of Camp Carson's Mountain and Cold Weather Training Command cross a chasm with a rope bridge during a summer demonstration. At left, they exhibit the evacuation of a casualty. The Army built the amphitheater for spectators to observe from below the cliffs as the soldiers demonstrated military mountaineering techniques in North Cheyenne Canyon. These free events had been offered to the public twice a week during the summer months since 1946. The purpose of these demonstrations was to teach the public about the mission of mountain-trained troops and to display the various skills they had acquired at the training command at Camp Hale. They hoped to help the public gain an appreciation of their Army.

The Mountain Command trained both summer and winter in a six-week course and were then integrated through the Army to pass on these valuable skills to others. In a real wartime situation, troop carrier planes delivered mountain-trained troops to inaccessible places by parachute drop. The soldiers in the canyon demonstrated how they would then scale sheer cliffs and rappel down vertical drops.

Nancy Thaden sits astride a friendly bay named Shadow in the early 1970s. Shadow's barn was located at Carson's Turkey Creek Ranch. Horseback riding at the ranch was a favorite activity, especially for this city girl raised back east. The dusty, red riding trails along dry creek beds and rocky, juniper bluffs became real-life landscapes she had only seen in old Western movies. (Angela Thaden Hahn.)

Many of the men who made up the cavalry platoon in this 1971 image had just returned from Vietnam. They trained during the morning hours at their assigned MOS with the mechanized troop of the 4th Squadron, 12th Armored Cavalry Regiment. In the afternoon, they volunteered at the Turkey Creek stables to rehearse their drills. The group began in the early 1960s as a four-

man mounted color guard and expanded to a platoon consisting of 20 men. Many of the men were ranch boys who wished to be around horses. They trained at Turkey Creek Ranch in historic cavalry formations such as the saber charge. (Courtesy of Jim Drye.)

The 4th Squadron of the 12th Cavalry Horse Platoon lines up in formation in 1971 at Turkey Creek Ranch. Astride their horses are, from left to right, Tom Bell, Gene Austby, Floyd H. Newcomb, John (Little John) Willis, Don Blair, Stanley W. Nolan, Murray Hagen, D. Douglas Biggs, L.M "Corky" Grayeagle, Carl Hinds, Robert G. McMillian, Robert D. Robinson, Mike Muscio, Paul Twichell, Roger Clark, and Kevin Johnson. Office in Charge (OIC) James E. Drye stands in front. They called themselves the Iron Horsemen and made it their mission to promote a favorable relationship between the public and the Army. The platoon received many requests from towns and civic groups to ride in their parades and perform mounted drills. The unit wore authentic uniforms and performed in a number of western states at rodeos, celebrations, and horse shows. (Courtesy of Jim Drye.)

Trying out for the Horse Platoon required good horsemanship. Tryouts involved saddling and riding one of the horses and putting the horse through paces, gallops, and turns. The young horse soldiers received their formational training from a former cavalry officer who taught them how to mount, form rank, and how to form the platoon in columns of twos and fours. Most of the horse soldiers lived in the former Spencer Penrose weekend home, which they used as a bunkhouse. In the center of this photograph are the long cavalry barn and the smaller dude barn. Turkey Creek meanders along the left. One former horseman tells of the time he and his partners were having some fun with their bunkhouse cook when they threw him in the creek on a freezing November day. Good-natured "Cookie" just laughed it off. (Courtesy of Jim Drye.)

People who came to the ranch to ride for a day or take horse riding lessons saddled up at the dude barn in the foreground. The cavalry used the longer barn. During a normal day, the men of the mounted horse platoon took care of their horses, cleaned the stalls, helped the ranch hands with some work, and spent several hours practicing mounted drills. (Courtesy of Jim Drye.)

About 30 horsemen lived in the bunkhouse. They were comprised of both the cavalry platoon and the dude barn crew. While the cavalry made many trips out of town for performances, the others took care of the ranch, baled hay, and shoed the horses. At the center of this 1971 image is ranch cook Richard Byrd with his signature smile that illustrated his typical, upbeat nature. (Courtesy of Jim Drye.)

OIC Jim Drye rode Cecil when working with the cavalry. He wears the historical cavalry uniform of the 1870s, which consists of a black slouch hat with a yellow cavalry cord and a brass insignia of crossed sabers. He wears a dark blue shirt with a yellow neckerchief and sky-blue trousers with yellow stripes along the side seams. A brown leather McClelland saddle sits atop Cecil's back. (Courtesy of Jim Drye.)

Nancy Drye takes her horse Narod out of his trailer for a break at her parents' home within Fort Carson's NCO quarters. She boarded her horse at the ranch where OIC Jim Drye lived. They enjoyed horseback riding together. When Jim exited the Army, he and Nancy were married and went to live in his hometown of Mesa, Arizona, with Narod and his horse, Cindy. (Courtesy of Jim Drye.)

Armed Forces Day is celebrated in May to acknowledge military men and women serving in the armed forces. It is usually recognized with parades, open houses, receptions, demonstrations, displays, and airshows. On this patriotic holiday, the military takes time to show the public various exhibits and displays to teach them about the country's military services. In the 1970s, personnel offered the public short rides in helicopters and armored personnel carriers (APCs). Exhibits displayed booby traps, land mines, and other battle hazards encountered by the United States and allied soldiers in Vietnam. In this photograph, curious visitors step inside a C-5 Galaxy airlifter to experience its cargo size. The Air Force commonly assisted the Army in transporting large, heavy equipment. This model had the capacity to easily transport the equivalent of six Greyhound busses. The C-5 had been in use since 1968 and was heavily utilized in Vietnam.

The 4th Infantry Division received its nickname, "the Ivy Division," from the Roman numeral four—IV. The shoulder patch contains a design of four ivy leaves, which are symbolic of tenacity and fidelity. These qualities are the basis of the division's motto: "Steadfast and Loyal." The division's second nickname, "Iron Horse," was later adopted to indicate the speed and power of the unit.

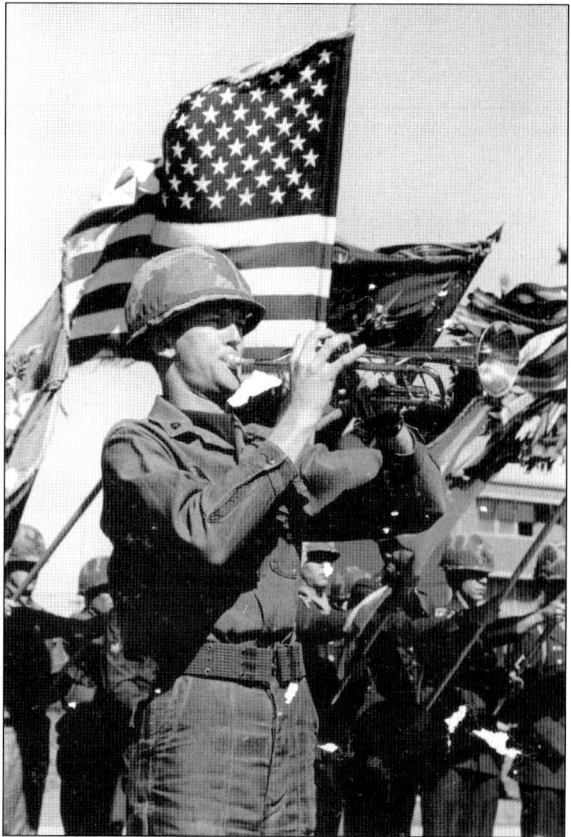

In Colorado Springs, the Army band makes another appearance in the community as it marches south on Tejon Street in a Pikes Peak or Bust Rodeo parade in the early 1970s. With the intersection of Colorado Avenue before them, the tall Exchange National Bank building looms in the background. The storefronts on the south side of Tejon in this image are still there.

Tank crews from the 6th Battalion, 32nd Armor Regiment armor battalion train at Fort Carson in the 1970s. Their Combat Vehicle Crewman (CVC) helmets and the tank camouflage pattern indicate their recent service in Vietnam. Each crew operates an M60 Patton tank, which was the Army's primary tracked vehicle. They attached large, bushy branches from native juniper trees to further camouflage their tanks. The armored vehicle came equipped with a 105mm gun. These tanks also included large, square searchlights, with infrared and normal lighting capabilities, to aid in nighttime movement. This reliable tank remained in service for more than 40 years.

A Carson tank crew takes an M41 Walker Bulldog tank out for some training. The Army mass-produced this tank in a hurry at the outbreak of the Korean War. It was not a popular model among American troops though, because the space in the hull was cramped. The United States sent a large number of these tanks to South Vietnam, where the smaller native soldiers found them adequate.

During tank training at Fort Carson in the 1970s, a crewman from the 4th Infantry Division communicates from a specially designed CVC helmet. This DH-132 model featured an outer shell over a soft liner. A microphone swiveled from the right earpiece. With his left hand, he operates the toggle switch to enable him to speak on the radio to personnel outside his immediate crew.

In the 1970s, a soldier from the 4th Infantry Division practices with an M29 81mm mortar on Fort Carson's training range. This portable weapon was used heavily during the Vietnam War. The soldier set the mortar upon a base plate and bipod, used the sight to make measurements, and loaded the muzzle with a high-arcing explosive shell.

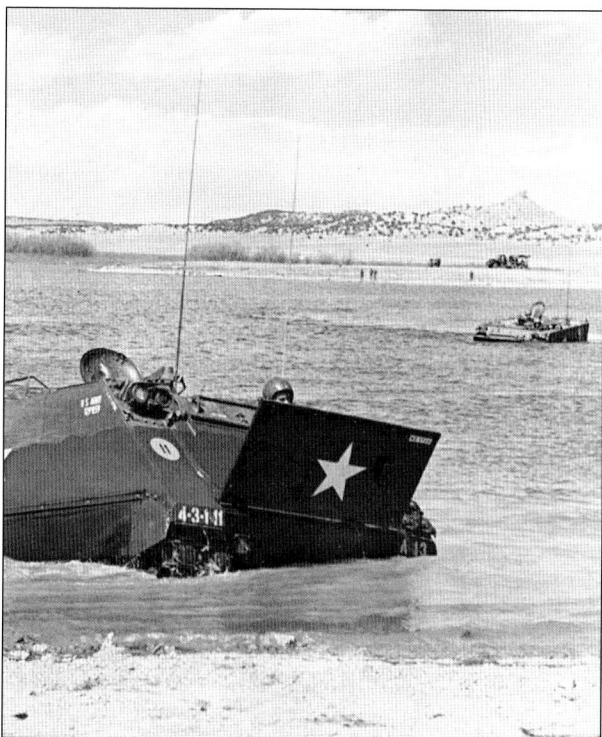

Soldiers in the 1970s take their M113 APCs for a swim in Carson's Teller Reservoir. A dam across Turkey Creek fills the lake, which became an excellent training spot for APC drivers. The new M113 saw heavy use during the Vietnam War, as the tracks on this light vehicle proved capable of charging through heavy jungle foliage and provided amphibious capability as well.

In the 1970s, recruits continued basic training at Fort Carson. Above, men practice moving over barbed wire, a mainstay of the military obstacle course. Part of their gear includes the M14 rifle, which proved to be a good weapon. However, due to its weight, it was put aside in favor of the lighter M16, though the M14 continued to see limited use. Below, trainees hang from monkey bars made for the big boys as they use their upper body strength to compete in being first across. The last unit of basic trainees graduated from Fort Carson in 1976.

A Fort Carson soldier receives instruction on the M60 machine gun, the Army's standard gun of its type since the Vietnam War. This weapon is usually operated by a three-man team—gunner, assistant gunner, and ammunition bearer. The gun and all its ammunition are so heavy, all three men carry a load of ammunition belts.

In the foreground, an M113 APC makes a pass down one of Carson's dusty tank roads as part of a convoy of other military vehicles headed downrange for maneuvers. In addition to personnel, APCs carried mortar tubes and medical evacuees. The ambulance jeep in the background waits for the convoy to pass.

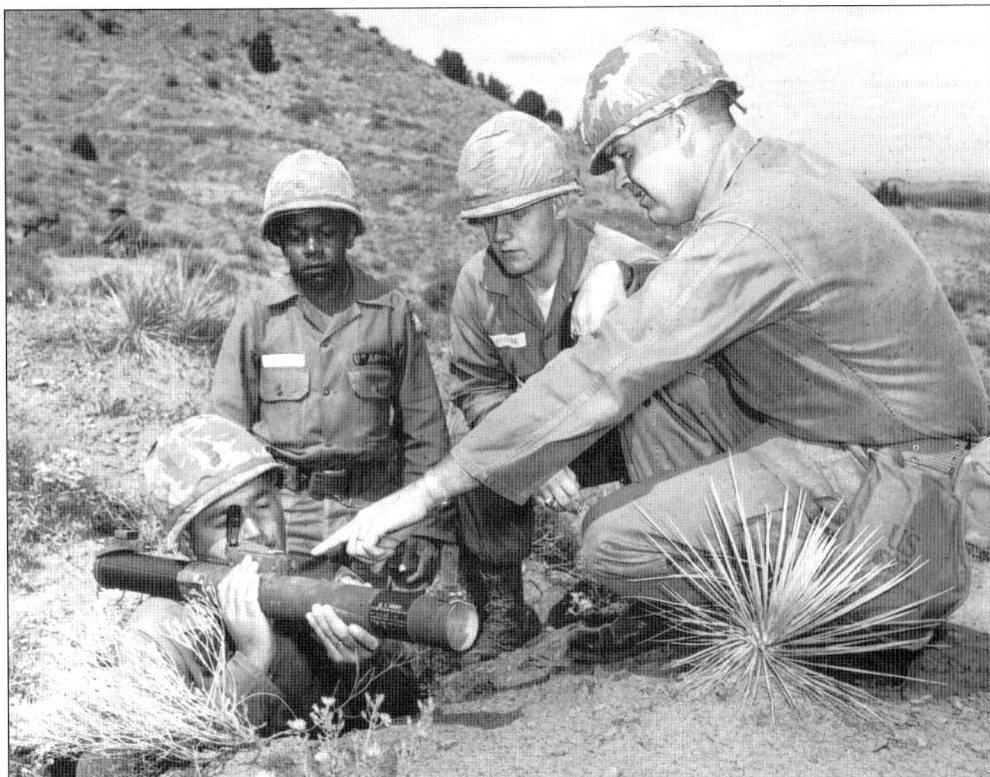

In the 1970s, a weapons instructor teaches Fort Carson men how to operate an M72 Light Anti-Armor Weapons System (LAWS) rocket launcher. The portable weapon propelled 66mm unguided missiles and became the preferred anti-tank weapon instead of the Super Bazooka. The miniaturized LAWS bazooka was lighter and did not require a two-man crew.

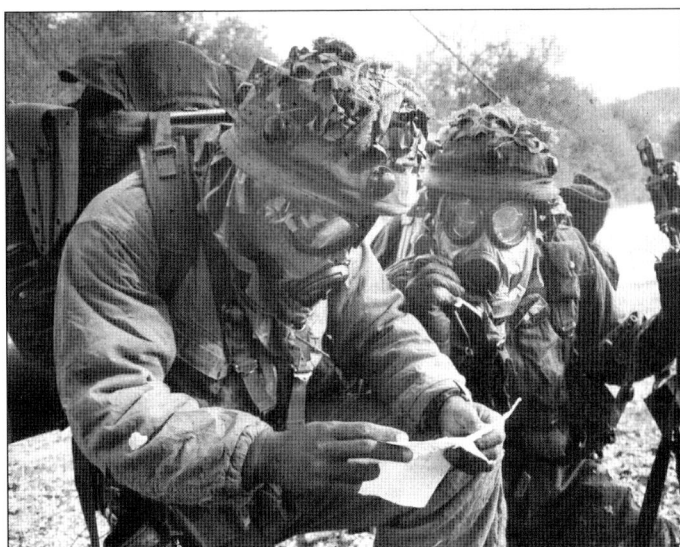

Soldiers participate in nuclear, biological, and chemical warfare training at Fort Carson during the Cold War. Instruction in this type of warfare began in the 1970s. Training in gas masks helped soldiers become accustomed to how they feel and how to function normally in them. Each gas mask kit came with a tube for drinking and decontamination chemicals.

Above, units of soldiers prepare for a military ceremony at Pershing Field despite snow on the ground and frigid temperatures. It took a considerable amount of time to get an entire division of men gathered and positioned in military formation. One veteran who was a corporal in the late 1970s remembers with fondness the typical hurry-up-and-wait routine. He said, "Imagine hundreds of snowballs soaring through the air above these men," when describing how they dealt with the waiting. Below, units of men move around the field in the military ceremonial protocol to recognize an important event.

The northern end of the motor pools appears in the foreground of this image. Beyond the large, brick barracks buildings, a mobile unit pushes forward in formation as they make their way around the parade field to the reviewing stands. Such divisional formations occurred at the change of a command or the activation or deactivation of a unit.

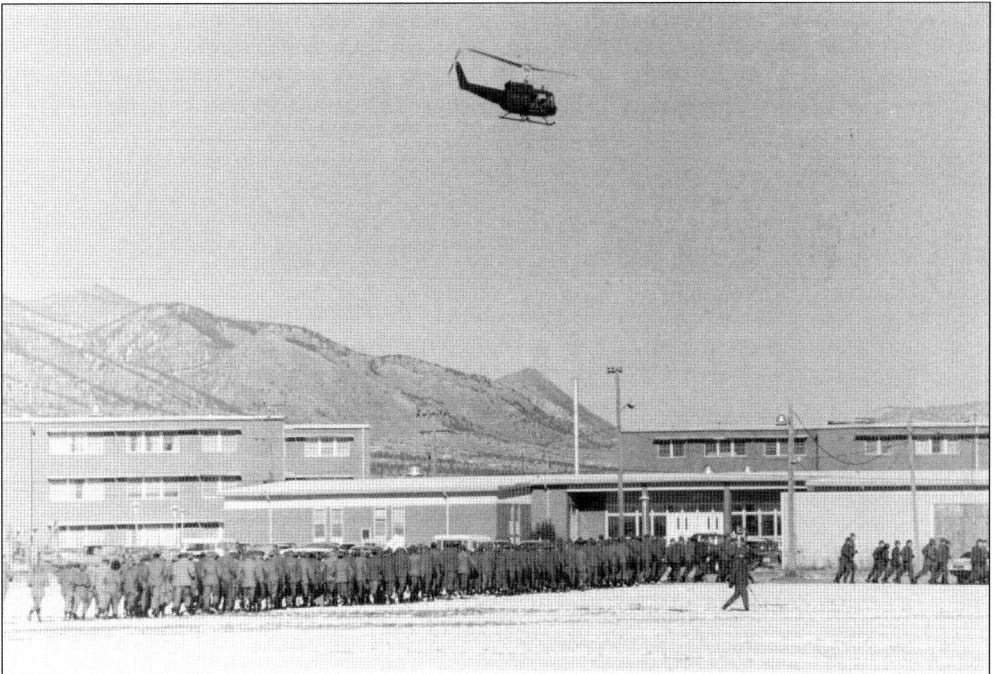

After formations on the field, the men head inside, and a "Huey" helicopter flies overhead. Sometimes, troops gathered for indoor training. Other times, they were required to attend an assembly or indoor ceremony. Huey is a nickname for the HU-1A helicopter that was developed in the 1960s and spent a considerable amount of time in Vietnam with the Army.

Still in Vietnam era camouflage, men from Fort Carson's 4th Infantry Division train on the new M1 Abrams tank in the 1980s. Later, it accompanied American troops to fight in the first Gulf War in 1990. The soldier kneeling in the foreground holds an M-60 machine gun, with a bipod folded against it, to use in firing from a prone position.

Sometime in the 1980s, one of Carson's commanding generals had many evergreens planted on post during his administration. The National Arbor Day Foundation bestowed the Tree City USA award to the Mountain Post for 22 consecutive years beginning in 1987. Carson was honored with the award for planting new trees, for transplanting trees from demolition sites, and for offering education in tree care.

Standing in front of the medevac helicopter is the Thaden family; from left to right are Nancy, Leslie Jr., Les, and Angela Hahn holding Maria Hahn. Special needs civilian Leslie Jr. and his family became the guests of medic Kim Strader at Butts Army Airfield in 1984. Strader provided Leslie Jr. with a closeup view of his favorite machine. After spending some time in the pilot seat, Leslie received his wings.

During the 1990s, Fort Carson's Turkey Creek Ranch offered a recreational spot for military personnel and their families. The ranch invited groups to utilize the picnic and camping grounds and offered various activities, such as the stagecoach ride in this image. Participants learned what it was like to ride in an old horse-drawn vehicle like the ones that passed by the ranch a century before. (Courtesy of Nancy Thaden.)

With the dude barn in the background, Nancy Thaden and her granddaughters enjoy a 1990s outing to Turkey Creek Ranch, where the girls get up close and personal with the resident horses. Her granddaughters are, from left to right, Jennifer Hahn, Maria Hahn, and Melissa Hahn. Also present in the photograph, but barely visible, is Melanie Hahn. The ranch provided these animals for riding lessons and trail rides into the ancient Turkey Creek drainage. They also carried members of the post's Mounted Color Guard. Additionally, the ranch offered boarding for horses belonging to anyone associated with Fort Carson. The Army benefitted a great deal from the purchase of this historic ranch. It came with tidy barns and outbuildings, well-established trees and grounds, and the delightfully attractive main house. (Courtesy of Nancy Thaden.)

Five

CONSTANT LIKE
GRANITE MOUNTAINS

In a gesture of support for its neighboring military post, the Cheyenne Mountain Garden Club presented this Blue Star Memorial to Fort Carson in 2005. It joins other memorials in the garden just outside the main gate. The National Garden Clubs began placing these markers in 1945 to pay tribute to the US armed forces. The blue star represents the star on flags displayed by families of combat soldiers.

Fort Carson MPs work to pull a Humvee from a snowbank on Fontaine Boulevard in the blizzard that struck the Pikes Peak region in October 1997. Eight soldiers in four Humvees responded to the disaster, venturing out to the eastern portion of El Paso County to assist local volunteers in rescue efforts. This storm is remembered as one of the worst in local history.

In November 2003, Pres. George W. Bush delivers a speech to soldiers and their families in a hangar at Butts Army Airfield. The soldiers of Fort Carson were then engaged in the largest deployment from the post since World War II in response to the war on terror, which began with the 2001 attack on the World Trade Center and Pentagon, and the thwarted attack on the White House. (Courtesy of the National Archives.)

During the Bridger Fire at Piñon Canyon Maneuver Site in 2008, a Fort Carson Fire Department brush truck conducts a burnout operation—in other words, fighting fire with fire. Carson and neighboring county firefighters train together and also spend time digging and conducting controlled burns to create fire breaks in preparation for possible future wildfires. (Courtesy of Fort Carson Fire Departments.)

Busloads of soldiers with the 4th Infantry Division pass under welcome home signs strung up on the Chadwick Street pedestrian bridge while heading west along Academy Boulevard to Fort Carson in February 2009. They had been away for a long 15-month deployment in support of Operation Iraqi Freedom. Starting at the end of January, about 3,800 troops from the 3rd Brigade Combat Team (BCT) returned over a one-month span. (Courtesy of the Department of Defense.)

The Fort Carson Mounted Color Guard bears the national colors while marching in formation at the front of the National Cavalry Competition parade. The guard consisted of 21 soldiers who traveled to Nebraska's Fort Robinson State Park in 2009. The Carson horsemen serve as ambassadors to the civilian communities in several western states, helping to promote good relations between the Army and its neighboring communities. They perform in historically accurate cavalry uniforms from the 1870s and participate in hundreds of events, ceremonies, and competitions each year. They are requested to highlight many parades, rodeos, holidays, state and county fairs, and heritage activities with their well-rehearsed drills. Even though horses have largely disappeared from standard military operations, the continued use of horses in ceremonies is part of Army tradition. At civilian events, the mounted color guard is very popular among spectators. (Courtesy of the Department of Defense.)

A team examines a Native American petroglyph found in a training area on Fort Carson in March 2010. Efforts began in 1980 to preserve tribal cultural and sacred sites from damage due to construction and training. Carson's Cultural Resources department consults with tribal representatives regarding the significance of historical sites, and soldiers receive information on how to train downrange while protecting tribal culture. (Courtesy of the Department of Defense.)

In August 2010, supply Sgt. Richard Sanford of the 1st BCT at Fort Carson calls cadence for his formation of men marching in the Red, White, and Brave Day parade in Colorado Springs. Mayor Rivera provided this opportunity for the community to show appreciation for all who serve and to particularly welcome home those who had recently returned from Iraq and Afghanistan. (Courtesy of the Department of Defense.)

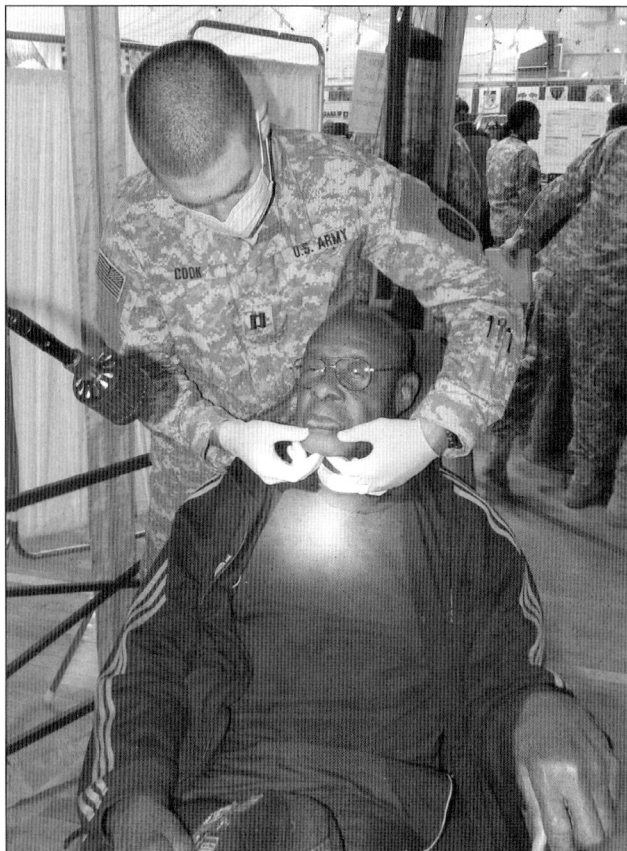

Members of the 10th Special Forces Group (Airborne) (SFG[A]) practice positioning and movement after exiting a helicopter. Although the 10th had been at Fort Carson since 1995, the group was excited to activate its 4th Battalion in August 2010. Special Forces learn cold-weather mountain warfare and advanced urban combat, among other skills. (Courtesy of the Department of Defense.)

In October 2012, Capt. Ross Cook, a dentist at Carson Dental Clinic, performs a dental screen for Ernest Martin at the 17th annual Retiree Appreciation Day. The event was held in the old gymnasium, now used as the SEV, for the benefit of military retirees and their widows. Specialists were on hand to conduct health screens and provide financial and legal advice. (Courtesy of the Department of Defense.)

Spec. Eric Joiner stands at parade rest with Master Sgt. Murphy at a retirement ceremony on Fort Carson's Manhart Field in August 2011. Murphy retired on this day with 13 other soldiers. After the men had received their certificates, Joiner led Murphy to the front, removed his saddle, and led the bay from the field. Brig. Gen. James H. Doty stood at the podium and announced, "Today we come together to honor the careers of 13 soldiers and one horse." This marked the first time in Fort Carson's history when a horse had been retired in a formal military retirement ceremony. Murphy served more than 11 years in the mounted color guard, participating in hundreds of ceremonies, parades, and other activities. He competed in several national cavalry competitions and won Top Cavalry Horse in 2005 at Fort Riley, Kansas. Murphy loved his duties as a cavalry horse but was forced into retirement due to his swayback condition. (Courtesy of the Department of Defense.)

Soldiers from the 759th Military Police Battalion direct a UH-60 Black Hawk helicopter from the 4th Combat Aviation Brigade (CAB), in a medical evacuation (Medevac) exercise to practice hoisting an injured soldier during a downed aircraft simulation on Fort Carson in August 2013. The reliable Black Hawk entered service in 1979 and has been used in assault operations as well. (Courtesy of the Department of Defense.)

Edith Nunez makes a pencil rubbing of her brother's name while her father, Isidro Nunez, looks on. They traveled from Texas to attend the 11th annual Mountain Post Warrior Memorial Ceremony held for the families of the fallen. Her brother's name, Staff Sgt. Joe Nunez, was one of nine added to the stone for the Memorial Day ceremony in 2014. The memorial is located in Kit Carson Park, just outside of Gate 1. (Courtesy of the Department of Defense.)

Above, aircrews from the 3rd Assault Helicopter Battalion within the 4th CAB and soldiers from the 3rd Armored Brigade Combat Team (ABCT) learn how to properly rig a sling on a package that is about to be sling-loaded by a Black Hawk on Fort Carson in September 2014. Although the 3rd is an assault battalion, they also learn sling-loading to assist in resupplying ground troops in unreachable areas. The sling-load operation provides a way for large, bulky objects to be transported by helicopter externally. Below, Sgt. Thomas Streb, a petroleum supply specialist, directs the aircrew over the package at his right. Almost all helicopters are capable of sling-loading and assist in this method of transporting supplies and vehicles when ground transportation is not an option. (Both, courtesy of the Department of Defense.)

Soldiers from the 1st Stryker BCT learn how to prepare and hook an M777A2 howitzer artillery piece safely and properly to a CH-47 Chinook from the 4th CAB on Fort Carson in August 2014. Sling-loading is a basic task for Chinook pilots, but training with a Humvee or heavy equipment like a howitzer is quite useful. Such training is essential to a unit's combat readiness in its ability to resupply troops on the front lines or in areas not suitable for vehicles. Combining both ground and aerial operations broaden the Army's capacity. (Both, courtesy of the Department of Defense.)

Above, an AH-64 Apache helicopter from the 4th CAB takes off from a forward arming and refueling point during an aerial gunnery range exercise on Fort Carson in September 2014. Below, an Apache fires simulation missiles in the same exercise. Aircrews prove their mission-readiness qualifications three to four times a year. They learn how to differentiate between hostile targets and friendly forces to avoid fratricide. They also learn how to provide necessary firepower to suppress the enemy and prevent them from overcoming ground troops. Apaches also provide an eye in the sky to watch over convoys and discourage ambushes. (Both, courtesy of the Department of Defense.)

Aircrews from the 4th CAB and soldiers from the 3rd ABCT conduct air assault operations on Fort Carson in September 2014. During this three-day training exercise, the two brigades practiced together in carrying supplies and troops around the battlefield. In addition to sling-loading, the troops practiced air assaults, in which aircraft move ground troops to terrain that has not been

fully secured and to places behind enemy lines. Honing communication between aircrews and ground forces is essential to wartime success. Air assault training during this particular exercise consisted of many attack rehearsals on Camp Red Devil. (Courtesy of the Department of Defense.)

Aircrews from the 4th CAB and soldiers from the 3rd ABCT take cover during a mock attack from 3rd ABCT's opponent force on Fort Carson in September 2014. Both brigades are from the 4th Infantry Division. Joint training gives combat soldiers from different MOSs the opportunity to learn to work together. (Courtesy of the Department of Defense.)

During a medevac response exercise in September 2014, the 4th CAB joins in training with the Fort Carson fire department, as a Black Hawk lowers a firefighter to a simulated downed aircraft in the mountains, to practice for possible aviation training accidents. (Courtesy of the Department of Defense.)

Downrange on Fort Carson, an M109 Paladin tank fires a 155mm missile from its howitzer. The Carson range was established before neighboring farms and ranch lands became acres of suburbs. With increasing complaints about the firing range, the post has implemented a noise management plan. Despite the criticism, many neighbors welcome the sound of freedom. (Courtesy of the Department of Defense.)

Soldiers of the 1st Stryker BCT move a "casualty" to an armored Stryker combat vehicle during a live-fire exercise on a cold day in January 2015. Stryker teams train to deploy swiftly and move through urban areas as well as open terrain. The Stryker, with its eight wheels, was designed to set out quickly like the light Humvee, yet still retain the heavy armor of a tracked tank.

Soldiers of the 1st Stryker BCT lift a heavy log over their heads 20 times while competing in the Ivy Heptathlon during Iron Horse Week in January 2015. The whole division comes together to demonstrate their teamwork and proficiency in numerous competitive events that involve a variety of battle tasks. The week also includes informational demonstrations and concerts. (Courtesy of the Department of Defense.)

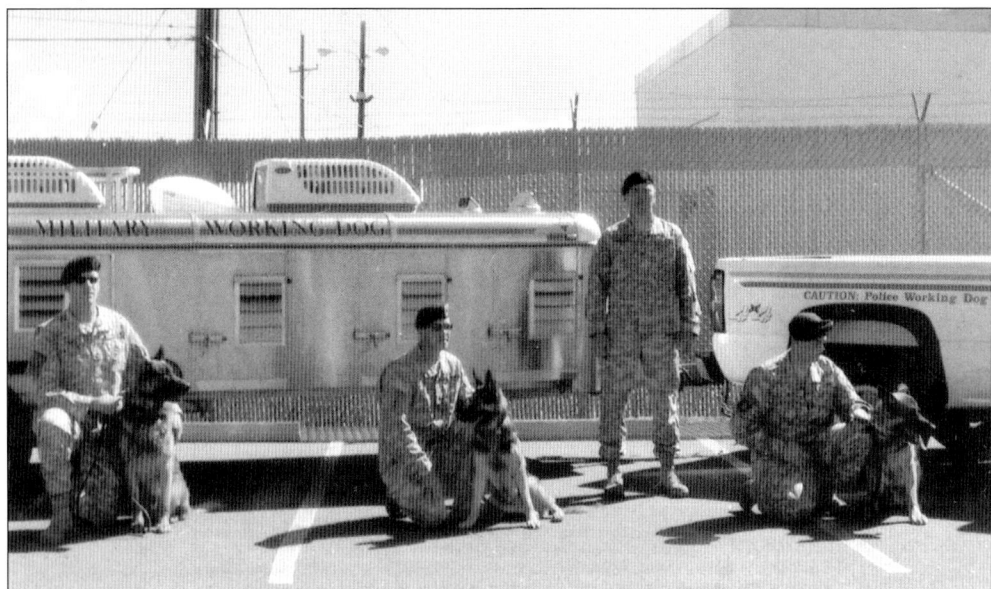

In July 2015, Fort Carson took its turn to host the annual week of certification testing for military working dogs. Military police handlers and their dogs came from Army posts throughout the country to prove themselves in their effectiveness in law enforcement and combat missions. The required testing covers areas of obedience, which include an obstacle course, law enforcement tactics, and the detection of drugs and explosives.

Soldiers of the 1st BCT, nicknamed the "Raider Brigade," participate in a march while training downrange on Fort Carson in April 2016. A Stryker team consists of soldiers and their Stryker vehicles who can be destructive, mobile, and durable. This particular Stryker brigade refer to themselves as the "first to fight." (Courtesy of the Department of Defense.)

These men participate in an emergency deployment readiness exercise with the Air Force. With camo-painted faces, Blackhawk soldiers of the 2nd Squadron, 1st Cavalry Regiment, 1st BCT, carry heavy weapons as they off-load from a C-130 Hercules aircraft at Peterson Air Force Base in April 2016. (Courtesy of the Department of Defense.)

Cpl. Manuel Rowland, a wheeled vehicle mechanic of the 1st BCT, prepares to throw a grenade onto a training bunker during a hand grenade qualification range in June 2016 on Fort Carson. Grenades are an intricate, highly explosive weapon that can hurl deadly fragments or a screen of thick smoke. (Courtesy of the Department of Defense.)

Capt. Elizabeth Mondo is one of the hundreds of women trained at Fort Carson. She started her medical career at Evans Hospital and searched for ways to deploy overseas. She was eventually attached to the 40th CAB in Camp Buehring, Kuwait, where in 2016 she served as a critical care nurse on a medevac helicopter. (Courtesy of the Department of the Defense.)

On an extremely hot, windy, and dry day in March 2018, live-fire training on Carson's range ignited a fire that quickly spread off the post to the neighboring community, forcing evacuations of the residents and their livestock. Faced with upcoming deployments, the decision was made to train on this day even though red flag conditions were extremely high for wildfires. Carson's firefighting team was at the training site as usual, but erratic winds abetted the fire, which literally exploded beyond their control across the combustible fuel of brown, dry grass. As the flames raced across the prairie like leaping fire demons, Army helicopters joined in the fight along with many other area fire departments. The Chinooks dipped their buckets into a nearby holding pond, then dumped the water near structures to discourage the flames. Returning residents were much relieved to find that the combined firefighting efforts saved all but two homes. (Courtesy of Michael Hahn.)

These two long, white buildings on Mekong Street, and three others, are all that remain of the old hospital. Once part of a busy complex of structures, personnel, and patients, they rest quietly now among shade trees. Old-timers remember that the hospital wards and its clinics were not fancy but were sufficient for transient soldiers and their families, as well as retirees, for over 40 years. (Angela Thaden Hahn.)

Kit Carson is depicted here as a young man at full gallop upon his horse. He grasps a Hawken plains rifle and wears a buckskin jacket. He carries a pouch for patches and balls, a powder horn, and a bedroll. The Friends of Fort Carson dedicated this bronze statue to the soldiers of Fort Carson in 2000 as a symbol of the relationship between the Army post and the local region. (Angela Thaden Hahn.)

About the 4th Infantry Division Museum

The museum displays artifacts as they relate to the US Army's 4th Infantry Division from its activation in 1917 to the present. The division has spent a considerable amount of time stationed at Fort Carson. The museum also hosts a yearly living history day in May.

The small museum is arranged as a walk through time, beginning with World War I and packing in artifacts, text, and visual scenery representing all 4th Infantry Division missions up to the present war on terror. The visitor learns of the Roosevelt connection to the division and of the 4th's direct involvement in the capture of Iraqi dictator Saddam Hussein.

The museum is located in building 6013, just outside Gate 1 of Fort Carson, on Nelson Boulevard and Highway 115 in Kit Carson Memorial Park. Current hours of operation are Monday through Saturday from 9:00 a.m. to 5:00 p.m. Call 719-524-0915 for more information.

DISCOVER THOUSANDS OF LOCAL HISTORY BOOKS FEATURING MILLIONS OF VINTAGE IMAGES

Arcadia Publishing, the leading local history publisher in the United States, is committed to making history accessible and meaningful through publishing books that celebrate and preserve the heritage of America's people and places.

Find more books like this at
www.arcadiapublishing.com

Search for your hometown history, your old stomping grounds, and even your favorite sports team.

Consistent with our mission to preserve history on a local level, this book was printed in South Carolina on American-made paper and manufactured entirely in the United States. Products carrying the accredited Forest Stewardship Council (FSC) label are printed on 100 percent FSC-certified paper.

MADE IN THE USA